OXFORD IB PREPARED

GEOGRAPHY

IB DIPLOMA PROGRAMME

Garrett Nagle

Anthony Gillett

OXFORD
UNIVERSITY PRESS

OXFORD
UNIVERSITY PRESS

Great Clarendon Street, Oxford, OX2 6DP, United Kingdom

Oxford University Press is a department of the University of Oxford. It furthers the University's objective of excellence in research, scholarship, and education by publishing worldwide. Oxford is a registered trade mark of Oxford University Press in the UK and in certain other countries

© Oxford University Press 2018

The moral rights of the authors have been asserted.

First published in 2019

Impression: 9

ISBN 978 0 19 843422 1

Printed and bound by CPI Group (UK) Ltd, Croydon, CR0 4YY

Acknowledgements

The publisher would like to thank the International Baccalaureate for their kind permission to adapt content from the subject guide.

The publisher and authors would like to thank the following for permission to use photographs and other copyright material:

Cover: robertharding/Alamy Stock Photo. All photos © Garrett Nagle, except: **p4**: Shutterstock; **p5**: Wikimedia Commons/CC BY-SA 3.0 IGO; **p13, 43, 47, 58: 76, 90, 92, 103**: Shutterstock; **p117**: iStockphoto; **p127**: National Weather Service/US GOV; **p146, 175, 186, 205, 206**: Shutterstock.

Artwork by Aptara Inc.

Every effort has been made to contact copyright holders of material reproduced in this book. Any omissions will be rectified in subsequent printings if notice is given to the publisher.

Contents

Answers to questions and exam papers in this book can be found on your free support website. Access the support website here:

www.oxfordsecondary.com/ib-prepared-support

This book provides coverage of the IB diploma syllabus in Geography and offers support to students preparing for their examinations. The book will help you revise the study material, learn the essential terms and concepts, strengthen your essay-writing and improve your approach to IB examinations. The book is packed with exam-style questions throughout to continually test your knowledge, and exam tips that demonstrate best practices and warn against common errors. All topics are illustrated by annotated example student answers to exam-style questions, which explain why marks may be scored or missed.

A complete set of IB-style examination papers at the end of this book provide further opportunities to check your knowledge and skills, boost your confidence and monitor the progress of your studies. Answers to all questions and examination papers are given online at **www.oxfordsecondary.com/ib-prepared-support**.

DP Geography assessment

All standard level (SL) and higher level (HL) students must complete the internal assessment and take papers 1 and 2 as part of their external assessment. Paper 1 examines the *geographic themes* and paper 2 examines the core *geographic perspectives*. Paper 3 is taken by HL students only, and examines three more geographic perspectives. The internal and external assessment marks are combined, as shown in the table below, to give your overall DP Geography grade, from 1 (lowest) to 7 (highest).

Overview of the book structure

The book is divided into several sections that cover the geographic themes and perspectives, internal assessment, and a complete set of practice examination papers.

The largest section of the book, the **geographic themes and perspectives**, follows the structure of the IB diploma geography syllabus (for first assessment 2019) and covers all *geographic inquiries* and *geographic knowledge and understandings*. At SL, you will study two options from the geographic themes and units 1–3 from the geographic perspectives. At HL, you will study three options and units 1–6.

The **internal assessment** section outlines the nature of the fieldwork that you will have to carry out and explains how to select a suitable topic, collect and process data, draw conclusions and present your report in a suitable format to satisfy the marking criteria and achieve the highest grade.

The final section contains IB-style **practice examination papers 1, 2 and 3**, written exclusively for this book. These papers will give you an opportunity to test yourself before the actual exam and at the same time provide additional practice problems for the material featured in all of the options and units.

The answers and solutions to all text questions and examination papers are given online at **www.oxfordsecondary.com/ib-prepared-support**.

Assessment overview

Assessment	Description	Units or options tested	SL		HL	
			marks	weight	marks	weight
Internal	Fieldwork with a written report	—	25	25%	25	20%
Paper 1	**Geographic themes** Structured and extended answer questions	Two options (SL) Three options (HL)	40	35%	60	35%
Paper 2	**Geographic perspectives—global change** Structured, visual and extended answer questions	Units 1–3 (Core)	50	40%	50	25%
Paper 3 (HL only)	**Geographic perspectives—global interactions** Extended answer questions	Units 4–6	—	—	28	20%

Key concepts

The **"Geography concepts"** model contains four key concepts—place, process, power, and possibility—and two "organizing" concepts—scale and spatial interaction. Scale has both temporal (short-term, long-term) and spatial (near, far) perspectives.

Places can be identified at a variety of scales, from local (for example, households, farms, villages) to the national or international level (countries, regions). Places can be compared according to their cultural or physical diversity, or disparities in wealth or resource endowment (for example, high income countries, low income countries, core-periphery). The characteristics of a place may be real or perceived (for example, migrants' views) and spatial interactions between places can be considered (for example, flows of people, goods and ideas).

Processes are human or physical mechanisms of change, such as transport and trade. They operate on varying spatial scales and timescales. Some processes may have negative impacts on people and the environment, whereas some may be more sustainable (such as circular flow processes).

Power is the ability to influence and have impacts at different scales. Unfair trading arrangements are a good example of a power divide. Power is may be concentrated in wealthy individuals, governments and institutions, and in processes in the natural world (for example, the power in hurricanes). Power has the ability to impact livelihoods, security, and cause unequal development.

Possibilities are the alternative outcomes that may occur. Key possibilities include human impact on climate change, the acidification of the oceans, degradation of farmland, population growth and the increased need for food, water and space.

The *concept link* feature, appearing throughout this book, allows you to connect the material of the geography syllabus to the four key concepts in this model.

Command terms

Command terms are the words in the exam question itself that tell you how to approach the question, and the detail and the depth expected. It is crucial that you interpret these terms correctly. The command terms will not be defined in any part of the exam papers; therefore, it is important to understand their meaning and their importance in advance of the exam.

Before you answer a question you should:

- Underline its command terms.
- Look at the mark weighting of that question.
- Match your answer to the depth required for the command term.

Term	Definition	Sample question	What you should cover
Analyse	Break down in order to bring out the essential elements or structure.	*Analyse the challenges associated with transboundary pollution.*	Describe and explain the social, economic and environmental issues associated with transboundary pollution.
Annotate	Add brief notes to a diagram or graph.	*Annotate the diagram to show short-wave radiation and long-wave radiation.*	Identify each of the types of radiation and add a clear label to show each one.
Classify	Arrange or order by class or category.	*Classify the following types of migration into forced- and voluntary-migrations.*	Decide whether each type of migration is likely to be undertaken voluntarily or whether it is forced.
Compare	Give an account of the similarities between two (or more) items or situations, referring to both (all) of them throughout.	*Compare the importance of wind and water in the development of landform features in hot, arid areas.*	You need to pick out the similarities—the impacts on erosion, transport and deposition, and the development of landforms.
Compare and contrast	Give an account of similarities and differences between two (or more) items or situations.	*Compare and contrast the ecological footprints that occur in HICs and LICs.*	For the both groups of countries, you should: • describe and explain the similarities (compare) in ecological footprints • describe and explain the differences (contrast)
Define	Give the precise meaning of a word, phrase, concept or physical quantity.	*Define the term "tourism".*	State a precise meaning of the term—there may be one or two marks be allocated to this question. Look at the mark weighting to determine the detail required.

Continued on next page

Term	Definition	Sample question	What you should cover
Describe	Give a detailed account.	*Describe two predicted trends shown on the graph.*	Look at the mark weighting to determine the detail required. You should adapt or manipulate the data provided to achieve full marks.
Determine	Obtain the only possible answer.	*Determine the month when the temperature range was greatest.*	Calculators are not allowed in the exam, and you may need to work out the value by eye or by using a ruler.
Discuss	Construct a considered and balanced review that includes a range of arguments, factors or hypotheses. Opinions or conclusions should be presented clearly and supported by appropriate evidence.	*"Population growth is the greatest threat to the earth's resources". Discuss this statement.*	Describe and explain different threats to the world's resources. You may decide that population growth is not a threat but a necessity to discover and develop more resources. Evaluate each view by drawing on case study evidence. Arrive at a conclusion that addresses different viewpoints but favours one more than the other.
Distinguish	Make clear the differences between two or more concepts or items.	*Distinguish between population distribution and density.*	Your answer should be more than two separate descriptions and it is essential that you emphasize the differences between them.
Draw	Represent by means of a labelled, accurate diagram or graph, using a pencil. A ruler (straight edge) should be used for straight lines.	*Draw a labelled diagram to show the impact of urbanization on a flood hydrograph.*	You should make sure that your diagrams are very distinct by drawing in black ink or pencil. Labels can be brief.
Estimate	Obtain an approximate value.	*Estimate the size of the lake in square 6327.*	You will have to use both the scale on the map and your ruler to obtain an approximate value.
Evaluate	Make an appraisal by weighing up the strengths and limitations.	*Evaluate the strategies to achieve sustainable urban development.*	Your evaluation is not just an opinion, but must provide evidence such as examples, case studies or facts to support it.
Examine	Consider an argument or concept in a way that uncovers the assumptions and interrelationships of the issue.	*Examine how inequalities between countries lead to different global flows.*	This means explaining thoroughly. It requires you to demonstrate in-depth understanding of both inequalities and global flows.
Explain	Give a detailed account including reasons or causes.	*Explain the causes and consequences of famine.*	You will need to briefly describe before you explain. You should explain a number of social, economic, political and environmental factors for both causes and consequences.
Identify	Provide an answer from a number of possibilities.	*Identify the direction of the most frequent wind.*	Only a brief answer (sometimes one word) is required here.
Justify	Give valid reasons or evidence to support an answer or conclusion.	*Methods of climate mitigation are shown in figure 1. Select two methods and suggest why they have the greatest potential to reduce carbon emissions. Justify your answer.*	You would need to explain the advantages and disadvantages of each methods as well as outlining the disadvantages of the methods not chosen.
Label	Add labels to a diagram.	*Label features A, B and C shown in the diagram.*	Each label is likely to be just one or two words. Do not describe or explain the feature.
Outline	Give a brief account or summary.	*Outline two ways in which a glacier erodes its load.*	This command carries few marks, so your answer should consist only of brief statements.
State	Give a specific name, value or other brief answer without explanation or calculation.	*State the landform shown in the photo.*	Only brief answers (sometimes one word) are required here.
Suggest	Propose a solution, hypothesis or other possible answer.	*Suggest reasons for the high level of infectious diseases in LICs.*	This term is used when there are several possible answers and you may have to give reasons or a judgement.
To what extent	Consider the merits or otherwise of an argument or concept. Opinions and conclusion should be presented clearly and supported with empirical evidence and sound arguments.	*To what extent are physical factors the main cause of water shortages?*	Your answer should consider physical factors along with human factors (social, economic and political) and the ways in which they are interrelated.

Many exam questions use terms such as describe and explain, but only the extended answer questions (essays) will expect you to discuss, evaluate and present an argument. Extended answer questions will include terms such as describe and explain, but also higher level commands such as examine, discuss, evaluate and to what extent.

These example questions illustrate the way in which a change of command produces a change in response. Each of the following is an extract indicating how you should respond, but not the complete answer.

Describe a population policy.

China introduced the one child policy in 1979. This was an anti-natalist policy which imposed financial penalties on couples having a second child and incentives for those upholding the law. The purpose of this policy was to curb high birth rates in an attempt to improve access to food and other resources, to reduce unemployment and to raise the national standard of living.

The policy reduced the birth rate from 33 per thousand in 1970 to 17 per thousand in 1979. In many rural areas, a preference for boys has led to a gender imbalance of 117:100 boys to girls. Fertility rates in many areas are well below replacement level. 2010 the rate was around 1.4, well below the replacement level of 2.1. As a result of the one child policy, China now has an ageing population. The policy was abolished in 2015.

Explain two impacts of the population policy.

China's one child policy was introduced in 1979 because rapid population growth was beginning to threaten economic progress. The policy was successful in reducing fertility rates and thus improving the livelihoods of families and reducing the economic burden on the state.

Enforcement in urban areas has been most marked and one child families have become typical. Traditionally, there is a preference for sons in Chinese culture, but a one child policy would inevitably lead to gender imbalance in the population because most families wanted a son to carry on traditions. Another consequence of the policy was an ageing population resulting from a fall in the fertility rate and greater life expectancy leading to an ever-increasing elderly population.

Evaluate the success or failure of a named population policy.

There are several adverse consequences of the policy. The preference for boys has created a gender imbalance. In the early days the policy was accused of contravening human rights. This is related to the abortion of female fetuses or female infanticide or abandonment. Both practices were practiced more so in rural than urban areas. The gender balance has created a shortage of child-bearing women now aged 20–40, which is socially undesirable and likely to reduce fertility rates still further. All this has negative economic implications. It will mean a future reduction in the labour force, lower revenue from taxation and the burden of an ageing society.

Exam guidance and strategies

Manage your time effectively: Every year, students lose marks through time mismanagement. The most common tendency is to spend too long on the first question at the expense of the others. Make sure that you are aware of the time allocation for each question and that you stick rigidly to it during the exam. Note that five minutes' reading time is allowed before the start of each exam.

Read the instructions carefully: Your revision should be thorough and no sub-topic should be omitted. For example, in paper 2 you will need to cover all three units of the core, but the questions will not necessarily cover all sub-topics within these core units. Familiarize yourself with the instructions at the top of your exam papers as these give you the instructions about timings and your choice of question.

Do not generalize: Geographers are keen on classification and putting phenomena into boxes. Generalization can sometimes be misleading and statements such as "China is a developing country" fails in several respects. First, China is too large in physical size and population to classify in this way. Second, levels of development vary greatly between rural and urban areas. Third, the rate of economic change is likely to make any classification obsolete within a few years. The solution is to choose a more reliable example such as smaller country at the bottom of the global economic development scale. Simplification can sometimes result in distortion of the truth. "Mumbai is a poor city" disguises the pockets of affluence and the existence of a very rich elite there. Generic expressions such as "e.g. Africa" used in exams reflect badly on the candidate.

Support your answers with evidence: Geography is about the real world, and in longer responses there must be plenty of factual support, examples and statistics.

Use current examples and case studies: There are a few unique and very well documented case studies that are frequently used in exams. One example is the Mississippi floods in 1993. But there are more recent and better documented examples of local floods. Outdated case study events are undesirable because human causes, consequences and responses have changed. Generally, the examples and case studies of geographic events occurring since 2000 are most relevant. Occasionally, older case studies, such as the Chernobyl disaster of 1986, are the best examples to use, in this case for a nuclear power disaster.

Use correct terminology: Good use of geographical terms shows understanding and avoids clumsy description. For example, a situation where "a population keeps on growing due to a lot of young people who are likely to keep the birth rate high for some years into the future", could be described as "population momentum".

Read the question carefully: For example, if you were asked to comment on the relative advantages of tourism as a development strategy in a less developed country and you discussed the issue with respect to a more developed country, you would inevitably lose marks. Ignoring commands is also common and is highly likely to lose marks.

The following table summarizes key information when sitting your exam:

Do	Don't
• Read the instructions on the cover of your exam paper to remind you of the exam regulations, such as the time allowed and the number of questions you should answer.	• Pad your answer with irrelevant content just to make it look better. Examiners are impressed by quality, not quantity.
• Underline the command terms in the questions and focus on these as you work through each question.	• Leave the examiner to draw conclusions if you cannot decide.
• Write a brief plan for essays, to give your answer a logical structure.	• Bend the question to fit your rehearsed answer.
• Observe the mark weighting of the sub-parts of structured questions.	• Spend too long on your best question at the expense of others.
• Give sufficient attention to the parts of the question requiring evaluation, discussion or analysis.	• Invent case studies; these will be checked by examiners.
• Complete the correct number of questions.	• Use lists or bullet points—these are not suitable for detailed analyses.
• Make sure that all your answers are legible, correctly numbered and in numerical order.	• Make your own abbreviations; for example, U for Urban and R for Rural. However, you can use accepted ones but make sure you write the term out in full the first time you use it—for example tropical rainforest (TRF) and infant mortality rate (IMR).

Key features of the book

Each chapter typically covers a Geographic theme or perspective, and starts with **"You should be able to show"** checklists. These outline the *geographic inquiries* and *knowledge and understandings* of the IB diploma Geography syllabus.

Chapters contain the features outlined on this page:

Test yourself

Test yourself boxes contain exam-style questions relating to the main text, where you can test your knowledge and understanding. The number of marks typically awarded to these questions is also given.

▶▶ Assessment tip

Assessment tips give advice to help you optimize your exam technique, warning against common errors and showing how to approach particular questions and command terms.

Question practice and sample student answer sections occur at the end of each chapter. This section includes typical IB-style questions relating to the chapter material, with advice on how best to approach these questions.

A sample student response to these questions is then given, with the correct points highlighted in green, and incorrect or incomplete answers are highlighted in red. Positive or negative feedback on the student's response is given in the green and red pull-out boxes.

An example of a question practice section and an accompanying student answer is shown below.

Key definitions are discussed at a level sufficient for answering typical examination questions. Most definitions are given in a grey side box like this one, and explained in the text.

🔗 Content link

Content links provide a reference to relevant material within another part of this book that relates to the text in question. Note that there are countless possibilities for linking content in the syllabus.

Concept link ⚭

Concept links connect the material of the geography syllabus to the four key concepts in the "geography concepts" model: places, processes, power and possibilities.

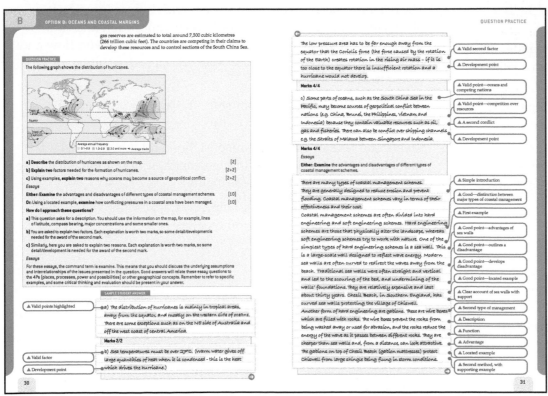

All of the questions in these sections have been written specifically for this book by the authors to reflect the new IB geography syllabus and examinations.

A FRESHWATER

A systems approach is a characteristic of the Geography course, and the Freshwater unit is no exception. A drainage basin is a system, since it has inputs, processes and outputs that shape landscapes and give rise to flooding. As water becomes scarcer due to human and natural factors, careful management of this precious resource is needed at local, national and international scales. As well as examining depletion, the pollution of freshwater is also studied. You should be able to make connections to other parts of the course as you progress through this unit, and you should ensure that you connect with the key concepts of processes, places, power and possibilities.

You should be able to show:

✓ how physical **processes** influence drainage basin systems and landforms;

✓ how physical and human factors exacerbate and mitigate flood risk for different **places**;

✓ the varying **power** of different factors in relation to water management issues;

✓ the future **possibilities** for management intervention in drainage basins.

A.1 DRAINAGE BASIN HYDROLOGY AND GEOMORPHOLOGY

- **Drainage basin** – an area of land that is drained by a river and its tributaries.

- **Watershed** – the border of a drainage basin that separates one drainage basin from another.

- **Open system** – when energy can enter and leave a system, such as a drainage basin.

- **Evapotranspiration** – the total amount of evaporation from land and from vegetation (transpiration).

- **Load** – the material transported by the river. The bed load consists of larger material that is transported via processes such as traction and saltation. The suspended load is transported via processes such as suspension and solution.

- **Cryosphere** – water in solid form (e.g. snow, ice).

You should be able to show how physical processes influence drainage basin systems and landforms:

✓ The drainage basin as an open system with inputs (precipitation of varying type and intensity), outputs (evaporation and transpiration), flows (infiltration, throughflow, overland flow and base flow) and stores (including vegetation, soil, aquifers and the cryosphere);

✓ River discharge and its relationship to stream flow, channel characteristics and hydraulic radius;

✓ River processes of erosion, transportation and deposition, and spatial and temporal factors influencing their operation, including channel characteristics and seasonality;

✓ The formation of typical river landforms, including waterfalls, floodplains, meanders, levees and deltas.

>> **Assessment tip**

This unit includes a wide range of terminology. Terms such as "eutrophication" and "salinization" are often not spelled correctly or are used out of context. Take time to practise spelling these terms, since appropriate use and spelling of terms will increase your mark for knowledge and understanding in your essay responses.

The drainage basin as an open system with inputs, outputs, flows and stores

A drainage basin is an open system since matter can enter and leave the system to join other systems such as a marine system.

After a period of rainfall, the water is then stored and transferred. Some of the transfers take place on the surface, such as overland flow (also known as surface run-off) which occurs when there is limited infiltration due to impermeable rock, for example. Infiltration occurs when water moves underground from the surface. A permeable rock type will allow water to pass through it, and the movement as it percolates downwards via gravity is classified as throughflow. Lakes, reservoirs, ponds, soil, vegetation and ice are all examples of stores where water is held. Clearly, there are a number of physical **processes** taking place in a drainage basin that affect the movement of water.

Test yourself

A.1 Distinguish between an open and a closed system. [2]

A.2 Analyse how a drainage basin functions. [3]

A.3 Explain how rock type and vegetation can affect the flow of water in a drainage basin. [2+2]

A.4 Study the map (figure A.1.1).

▼ Figure A.1.1. River Seine drainage basin

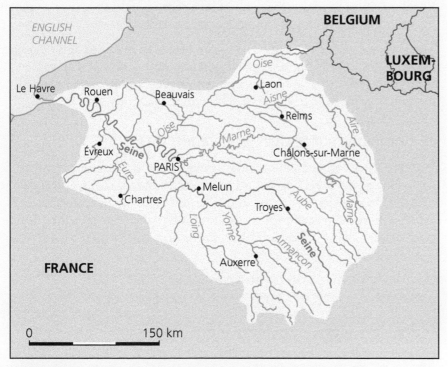

Describe the physical characteristics of the River Seine and its drainage basin. [3]

Concept link

PROCESSES: This section outlines a range of natural processes which create an open system. The dynamic nature of these processes not only shapes landscapes, but it also creates the unique characteristics of places which can be at a range of different scales, from the Nile basin to a basin at a local scale.

≫ Assessment tip

Make sure that you do not include "opposites" or "mirrors" in your answer. This is when you state the opposite compared to the previous part of your answer. You will only get a mark for one side of the "mirror", so do not expect double marks.

≫ Assessment tip

When asked to describe a map, it is important that you utilize your cartographic knowledge by including distances, compass directions, grid references and relief.

River discharge and its relationship to stream flow, channel characteristics and hydraulic radius

The discharge is the volume of water that is flowing at a given point in the river. It is measured in cubic metres per second (cumecs) and theoretically increases downstream. It is calculated by multiplying the velocity by the cross-sectional area at a point in the river. Wider and deeper channels will be able to hold a greater volume of water resulting in a larger hydraulic radius, or in other words, a higher efficiency or ability for water to move downstream.

River processes of erosion, transportation and deposition, and spatial and temporal factors influencing their operation

There are four processes of erosion: hydraulic action, corrasion, corrosion and attrition. Once material has been eroded, it is then transported down the river either as the bed load (material on the riverbed) or as the suspended load (material held in suspension by the flow of the water) or as the dissolved load (soluble material dissolved in the water). Some material may be carried on the surface, such as leaves and branches.

There are four processes for transportation: traction, saltation, suspension and solution.

When material is no longer being transported by a river, it is deposited. For deposition to occur, there must be a reduction in the river's velocity in order that the material can no longer be carried.

The seasonal nature of some rivers can mean an increase or a decrease in processes of erosion, transportation and deposition at different times of the year. For example, ephemeral rivers only contain flow at intermittent periods of the year when there is a rainy season.

The formation of typical river landforms

Waterfalls are formed when water flows over two different types of rock. One of the layers is eroded more easily than the other, and due to this the more resistant rock is undercut. Eventually the undercut rock is left without support underneath and it will collapse into the plunge pool below. The process continues, and a gorge is formed. A waterfall is a landform created via erosion.

▼ Figure A.1.2. The Goðafoss waterfall in Iceland

Processes of erosion, transportation and deposition are all required for the formation and evolution of meanders. Material is deposited on the inside and lateral erosion occurs on the outside of the meander.

When the sinuosity or "bendiness" of a meander increases, it is possible that the outside bends of two meanders are eroded in order that the meander straightens, creating an oxbow lake. See figure A.1.3 which shows the Juruá River, one of the longest tributaries of the Amazon River. It is a very sinuous river and the image shows a number of oxbow lakes.

▼ Figure A.1.3. Oxbow lakes on the Juruá

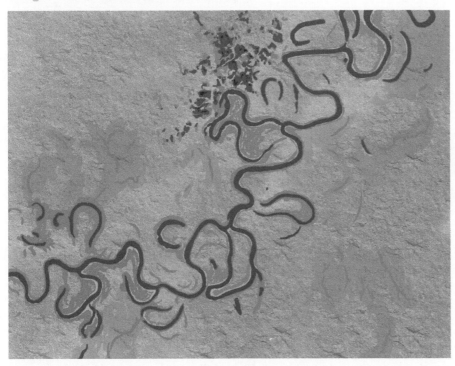

A floodplain is formed via the deposition of material adjacent to a river. When a river floods on the flat land at the side of the river (normally in the lower course), material is deposited once the water subsides. Layer upon layer of deposited material such as silt or alluvium further develops a floodplain. Levees are created during times of flood when the larger, coarser material is deposited closer to the river channel, as the amount of energy decreases with distance from the channel. Floodplains and levees are depositional landforms although the processes of erosion and transportation are needed to supply the material that is then deposited.

There are different types of deltas: arcuate, bird's foot and cuspate, and each is created via deposition.

Test yourself

A.8 (a) State two landforms found in a drainage basin that are formed solely via processes of erosion. [2]

(b) Explain how the landforms you identified in part (a) are formed. [2+2]

>> **Assessment tip**

A clear annotated diagram is an appropriate approach to answering this question.

A.2 FLOODING AND FLOOD MITIGATION

- **Hydrograph** – a graph that shows how a river or stream's discharge changes over time and its relationship with the amount of precipitation that falls during a rainfall event.

- **Antecedent moisture** – the amount of moisture stored underground after a previous period of precipitation.

- **Peak discharge** – the greatest amount of discharge flowing in a river after a rainfall event.

- **Peak rainfall** – the time at which there is the highest amount of rainfall into a drainage basin for a given storm.

- **Afforestation** – the process of planting trees in an area where there were previously none.

- **Reforestation** – replanting trees in an area that was previously deforested.

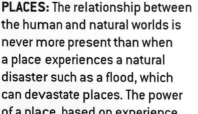

Concept link

PLACES: The relationship between the human and natural worlds is never more present than when a place experiences a natural disaster such as a flood, which can devastate places. The power of a place, based on experience and/or the economic ability to put in place defences to mitigate the effects of flooding, is able to ensure that the relationship does not bring negative consequences.

You should be able to show how physical and human factors exacerbate and mitigate flood risk for different places:

✓ Hydrograph characteristics (lag time, peak discharge, base flow) and natural influences on hydrographs, including geology and seasonality;

✓ How urbanization, deforestation and channel modifications affect flood risk within a drainage basin, including its distribution, frequency and magnitude;

✓ Attempts at flood prediction, including changes in weather forecasting and uncertainty in climate modelling;

✓ Flood mitigation, including structural measures (dams, afforestation, channel modification and levee strengthening) and planning (personal insurance and flood preparation, and flood warning technology);

✔ Two contrasting detailed examples of flood mitigation of drainage basins.

Hydrograph characteristics and natural influences on hydrographs

A hydrograph shows data for two variables on the same chart, e.g. a river's discharge and the amount of rainfall that a drainage basin receives. The amount of rainfall influences the amount of discharge, and this is via overland flow, throughflow and groundwater flow. The lag time—the time between peak rainfall to a river's peak discharge— is an important period. A short lag time means that a river may reach its bank full discharge, which is the maximum amount of discharge a channel can hold, and flood an area quickly. Increasing the lag time and reducing the discharge can reduce the risk of flooding. The monitoring of hydrographs can enable predictions to be made regarding flooding as well as measuring the effectiveness of flood mitigation strategies.

Test yourself

A.9 Discuss how physical factors can influence a hydrograph.   [3+3]

» Assessment tip

If you are asked to discuss the factors that influence a hydrograph in an extended response, it is important that you evaluate the impact, since this is normally necessary to reach the highest mark bands. For example, human activities can decrease the discharge and increase the lag time via afforestation, whilst urbanization and building on a floodplain will increase discharge and reduce the lag time.

A.10 Suggest how hydrographs can be used to forecast and manage flooding.   [2+2]

>> **Assessment tip**

When discussing hydrographs in your answer, it is appropriate to include an annotated hydrograph showing the different responses from a river when there are natural differences between two drainage basins or there is change annually in a particular basin.

How urbanization, deforestation and channel modifications affect flood risk within a drainage basin, including its distribution, frequency and magnitude

Human factors such as urbanization will mean that there are more impermeable surfaces, thus increasing overland flow and the time taken to reach a nearby stream or river.

Deforestation reduces interception and exposes soil to erosion whilst increasing the amount of overland flow due to soil capacity being reached.

Increasing the cross-sectional area of a channel reduces the risk of flooding since the channel can hold more volume.

The frequency and magnitude of a flood will depend on the natural and human factors that influence the movement of water. Infrequent flood events tend to have a large magnitude and vice versa. In summary, the characteristics of places via a range of human and physical factors will influence the risk of flooding.

▼ Figure A.2.1. Urbanization means an increase in impermeable surfaces and a reduction in infiltration

Attempts at flood prediction

Weather forecasting can predict the timing and the amount of rainfall during a period of low pressure and this data will forewarn authorities and residents in a drainage basin about the impending risk of flooding. Extreme weather events such as cyclones will increase the amount of rainfall and debris reaching a river. Meteorologists are able to detect and track these events days, possibly a week, before they arrive in an area.

Flood mitigation

There are a variety of flood mitigation strategies, and they can be classified as "hard" or "soft" engineering. Hard engineering is when the structural measure uses artificial materials with which to alter the landscape. This may involve building a concrete dam in order to control the flow of a river, or construct levees to form channels that will hold more water and protect nearby infrastructure. Soft engineering works with nature, so no concrete or artificial structures are created. Afforestation and reforestation would be examples of this type of flood mitigation strategy. Other strategies such as widening, deepening, lining and straightening a channel (channelization) will also enable a river to hold more discharge and increase its hydraulic radius, thus becoming more efficient.

Two contrasting detailed examples of flood mitigation of drainage basins

The following case studies (Pakistan and Queensland) show that although both places had flood management strategies in place, there was still significant economic damage and the loss of life.

Case study: Flood mitigation in Pakistan

▼ Figure A.2.2. The Indus river system

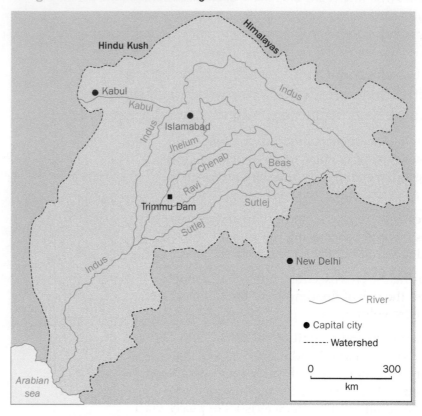

Pakistan has regular floods due to heavy monsoon rainfall, snowmelt from the Himalayas and deforestation.

In September 2014, the rivers Jhelum, Chenab, Ravi and Sutlej started rising above their banks due to monsoon rainfall, and they flooded. Around 2.5 million people were reported to have been affected; 367 people were killed and over 125,000 homes were destroyed. Warnings about the flooding came too late for many people.

Additional dams had been built on rivers further upstream in neighbouring India and some suggested that the release of floodwaters from these dams increased the discharge and amount of flooding further downstream. The Pakistani government denied this. After spending millions on barrages and embankments, the authorities had to allow controlled breaches of the embankments such as on the west bank of the Chenab upstream from a major dam at Trimmu, in order to protect urban areas. This meant that 200 villages were flooded.

Case study: Flood mitigation in Queensland, Australia

In January 2011 there was a major flood in Queensland, Australia due to higher than normal rainfall in a La Niña year and also from Tropical Cyclone Tasha. Rivers such as the Burnett topped their banks and flooded large portions of the state. The economic cost of the disaster was approximately A$10 billion and 35 people died.

Flood management was in place, such as the Wivenhoe Dam which was built to stop Brisbane being flooded (and also to secure water supplies). Residents argue that insufficient water was released from behind the dam in spite of forecasts of heavy rainfall. When water was eventually released, the Brisbane River already had a high discharge due to rainfall amounts, so the level rose by 10 metres causing widespread damage.

Test yourself

A.11 Using evidence from figure A.2.2, **outline** the potential difficulties for the Pakistani government in relation to mitigating flooding. [3]

A.12 Evaluate the success of flood mitigation strategies in Pakistan from 2010 onwards. [4]

> **Content link**
> Flood mitigation in cities is explored in option G.4.

A.3 WATER SCARCITY AND WATER QUALITY

You should be able to show the varying power of different factors in relation to water management issues:

✓ Physical and economic water scarcity, and the factors that control these including the causes and impacts of droughts; the distinction between water quantity and water quality;

✓ Environmental consequences of agricultural activities on water quality, to include pollution (eutrophication) and irrigation (salinization);

  ✔ Detailed examples to illustrate the role of different stakeholders;

✓ Growing human pressures on lakes and aquifers, including economic growth and population migration;

✓ Internationally shared water resources as a source of conflict;

  ✔ Case study of one internationally shared water resource and the role of different stakeholders in attempting to find a resolution.

Physical and economic water scarcity—water quality and quantity

Water stress is when annual water supplies drop to less than 1,700 m³ per person per year, whereas water scarcity is when a person has access to less than 1,000 m³ annually. The causes of water scarcity can be natural and human, for example a lack of rainfall in an area for a significant period and the unsustainable consumption of water by agriculture.

A decrease in water quality can be caused by a number of different factors. For example, the disposal of plastic in the world's oceans and rivers has led to fibres being present in the food chain and in drinking water. Chemicals from commercial agricultural are another significant source of contamination (see following page), while inadequate sewage and sanitation systems may pollute clean water. Waterborne diseases have caused the deaths of millions, and more than 3 million deaths a year are due to diseases such as cholera, malaria and diarrhoea.

> • **Physical water scarcity** — where water resource development is approaching or has exceeded unsustainable levels; it relates water availability to water demand and implies that arid areas are not necessarily water scarce.
>
> • **Economic water scarcity** — where water is available locally but not accessible due to human, institutional or financial capital reasons.
>
> • **Drought** — a prolonged period of abnormally low rainfall. Drought is a broad category and can be subdivided into hydrological, meteorological and agricultural drought.
>
> • **Eutrophication** — a process that affects freshwater whereby dense algal and plant growth occurs due to increased concentration of chemical nutrients.
>
> • **Salinization** — a process in which the salt content of surface and/or groundwater increases as overland flow or throughflow transfers crystallized salts left behind after irrigated water has evaporated.

> **Content link**
> Water security is examined further in unit 3.2.

Environmental consequences of agricultural activities on water quality

Agriculture can have a negative impact on freshwater. The use of irrigation and the application of chemicals in the form of fertilizer and pesticides can pollute freshwater on the surface and underground. Algae grow on the surface of rivers due to eutrophication which reduces the amount of oxygen available for vegetation and creatures under the water.

A reduction in oxygen results in a reduction of life in a lake or river. A number of different groups and organizations are trying to resolve this problem, including farmers (subsistence or commercial), environmental groups, residents and government departments such as the Environment Agency in the UK.

Case study: The role of different stakeholders in North Carolina, USA

Jordan Lake suffers from eutrophication due to nutrient rich run-off into tributaries that feed into the lake. A number of different stakeholders are involved in this issue, such as the state government which has the power to provide funding for the clean-up of the lake and the national government which is able to put in place rules for industry and agriculture. Here are some of the stakeholders:

- 300,000 people who rely on the lake for drinking water
- The Environmental Protection Agency (EPA), which enforces legislation from the Clean Water Act
- The state government of North Carolina, which provides funding for cleaning the lake
- The Army Corp of engineers who maintain the lake and dam
- Clean Jordan Lake is an environmental organization that organizes volunteers to help clean the lake
- 30 animal-feeding operations
- 12 wastewater treatment facilities.

Content link

The environmental impact of global agribusiness is explored further in unit 6.2.

Concept link

POWER: As water becomes more scarce due to the effects of global climate change, population growth and increased consumerism, the opportunity to access water can be decided by the power of a place. The places upstream in a drainage basin have the upper hand in controlling the amount of water to reach places further along the course of a river. In addition, the quality of the water can decrease due to agricultural and industrial processes, which can also negatively affect places.

▼ Figure A.3.1. **Distribution of water scarcity**

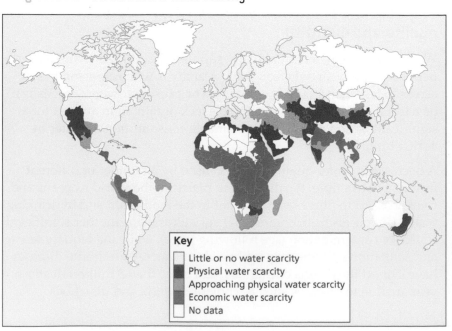

Key
- Little or no water scarcity
- Physical water scarcity
- Approaching physical water scarcity
- Economic water scarcity
- No data

Test yourself

A.13 Distinguish between physical and economic water scarcity. [2]

A.14 Study figure A.3.1, which shows the distribution for different types of water scarcity.

Describe the pattern of economic water scarcity. [3]

A.15 Examine how agriculture can have a negative impact on freshwater. [2+2+2]

Growing human pressures on lakes and aquifers, including economic growth and population migration

Aquifers (rocks that contain significant quantities of water) are not subject to evaporation, but they are at risk of significant depletion due to population increase via migration and also from economic growth. Economic growth can create pressure as consumption increases due to industrial processes requiring greater amounts of water as an input. In addition, economic growth may result in an increase in disposable incomes and the growth of a country's middle-class. A person's diet may change, with increased consumption of food, which leads to increased agricultural production and increased stress on water sources.

Groundwater is an important source of freshwater around the world. Aquifers can be either unconfined (they can be recharged via percolation) or confined (the water is enclosed between layers of impermeable rock). For confined aquifers, the recharge has to take place artificially since percolation is not possible.

Internationally shared water resources as a source of conflict

The political borders of countries and the watersheds of drainage basins are never a perfect fit. Many drainage basins straddle two or more countries. The residents of those countries may rely on the same freshwater source, on the surface and/or underground, to fulfill their needs. Agreements are often needed in order that water is shared equitably and consumption by the different stakeholders is environmentally sustainable. This becomes increasingly important as economic development takes place and populations increase.

Case study: Internationally shared water resource

Over recent decades there has been tension between the 11 countries that share the waters of the River Nile. Uganda, Egypt, the Republic of the Sudan, South Sudan, Rwanda, Burundi, Ethiopia, Kenya, Tanzania, Eritrea and the DRC all have a vested interest in the Nile since the drainage basin falls within each country's borders (it covers 10% of Africa's landmass). Egypt has a long-established historic agreement signed in 1929 and amended in 1959, which gives it (and the Republic of the Sudan) full access to water from the Nile, its only source of water, and an ability to veto any plans to restrict water upstream, such as the building of a dam.

>> **Assessment tip**

For short-response questions, try to avoid writing a long list of country names when you are asked to describe data shown on a map. As a geographer, you should be skilled at identifying general patterns and anomalies. You will only receive 1 mark at most for a 3-mark question if you provide a list. A 3-mark question will require three distinct patterns when describing a map. A useful strategy would be to comment on the pattern, the extremes and any anomalies.

>> **Assessment tip**

Be aware that the syllabus specifically requires you to connect a *human reason* (such as irrigation) with salinization.

In order to generate hydroelectric power for the country and to export it to other countries, Ethiopia decided to build the Grand Ethiopian Renaissance Dam (GERD) on the Blue Nile (the White and Blue Nile join to form the River Nile) and construction began in 2011. Egypt was immediately concerned since it expected a 25% reduction in the water that it would normally receive. At one point, a former Egyptian president proposed military action against Ethiopia.

Despite this past friction, in March 2015, Egypt, the Republic of the Sudan and Ethiopia signed a Declaration of Principles agreement. Part of the agreement was that an independent assessment would take place to evaluate the impact of the dam and ensure that each country would not be affected detrimentally. This independent assessment has never been published due to issues around getting access to accurate information from each country. Therefore, tension remains between the three countries, especially when other issues are considered, such as reduced rainfall and a growing population in Nile basin countries.

Test yourself

A.16 Suggest possible challenges for countries that have to share a source of freshwater.   [2+2]

A.4 WATER MANAGEMENT FUTURES

- **Integrated drainage basin management (IDBM)** – a comprehensive approach to the planning and management of a drainage basin involving a variety of different stakeholders in order that there is a balance between economic development and environmental impact.

- **Wetlands** – areas of marsh, fen, peatland or water, whether natural or artificial, permanent or temporary, with water that is static or flowing, fresh, brackish or salt.

You should be able to show examples of future possibilities for management intervention in drainage basins:

✔ The importance of strengthening participation of local communities to improve water management in different economic development contexts, including sustainable water use and efficiency, and ensuring access to clean, safe and affordable water;

✔ Increased dam building for multipurpose water schemes, and their costs and benefits;
  ✔ Case study of contemporary dam building expansion in one major drainage basin;

✔ The growing importance of integrated drainage basin management (IDBM) plans, and the costs and benefits they bring;

✔ Growing pressures on major wetlands and efforts to protect them, such as the Ramsar Convention;
  ✔ Case study of the future possibilities for one wetland area.

Concept link

POSSIBILITIES: Possibilities from managing freshwater within a drainage basin can bring benefits and problems at a variety of scales. The implementation of IDBM aims to bring benefits to all stakeholders in a basin, whether they live in different countries or have different priorities.

Strengthening participation of local communities to improve sustainable water use

The sustainable use of water at a national and international level has already been discussed in this chapter. Local communities can also improve the management of their water supplies by recycling water for residential gardens or using more efficient irrigation methods, such as drip irrigation pioneered in Israeli communities, to ensure that no water is wasted when growing crops. Empowering local people is another benefit of community involvement in safeguarding water supplies. The Self-Employed Women's Association (SEWA) in India is just one example of how women are trained to repair hand pumps in order that rural areas can secure clean water without having to walk several miles to an alternative source.

Increased dam building for multipurpose water schemes

Multipurpose dam schemes can provide protection against flooding and an area for recreation and fishing. They can also generate significant amounts of energy in a manner that has fewer emissions than the continual burning of fossil fuels. However, they can be detrimental to the natural environment. The difference in temperature between the water downstream and the water released from the reservoir causes problems for fish, while evaporation and siltation can reduce the amount of water in the reservoir and the potential for generating energy. The construction of dams releases large amounts of carbon dioxide due to the amount of concrete required. The money used to build a dam could also be used instead to significantly improve many aspects of a country, such as health and education.

Case study: Contemporary dam building expansion—the Lesotho Highlands Water Project (LHWP)

The Lesotho Highlands Water Project (LHWP) is a complex network of pipelines, tunnels and dams which divert water from Lesotho to the Gauteng region in South Africa (cities such as Johannesburg, Pretoria and Vereeniging) and the dams generate hydroelectric power for Lesotho. Lesotho has a surplus of water and the whole country is enclosed within the Orange river drainage basin. Once fully completed, the LHWP will have five dams with 2,000 million cubic metres of water transferred through 200 km of tunnels from Lesotho to South Africa every year. This is a significant engineering project with three phases of construction. The first was completed in 2004 and the second is to be finished in 2025, with no date given for the third phase. For Phase 1, approximately 4,857 hectares of arable land were flooded and 3,400 households were relocated to create the reservoirs necessary behind the Mahole and Katse dams. The Katse Dam was opened in 1998 and then the Mahole Dam in 2004. The Polihali Dam will be completed by 2025 as part of Phase 2.

▼ Figure A.4.1. The Katse Dam in Lesotho

The benefits for Lesotho were 16,000 jobs created during Phase 1, with 3,000 more jobs in Phase 2, over US$500 million in water sales and US$71 million from the sale of electricity by 2016. South Africa received much cleaner water which did not require treatment. The costs were that 20,000 people were displaced during Phase 1, and 17 more villages will be relocated during Phase 2. Inadequate compensation was provided to the people, and bribery was uncovered involving 12 construction companies during Phase 1 with allegations of malpractice during Phase 2.

The growing importance of integrated drainage basin management (IDBM) plans, and the costs and benefits they bring

As you have already seen a number of times in this chapter, growing populations, economic development and global climate change bring challenges when trying to please different stakeholders economically and socially, while also aiming to achieve environmental sustainability. IDBM tries to achieve all of these with the mutual cooperation of different stakeholders and different countries.

Test yourself

A.17 Suggest reasons why the benefits gained from the construction of large dams may outweigh any costs.   [3+3]

A.18 Briefly **outline** what is meant by the term IDBM.   [2]

A.19 Describe an IDBM using an example that you have studied.   [3]

A.20 Suggest how an IDBM can bring both costs and benefits to different stakeholders.   [3+3]

▼ Figure A.4.2. Wetlands in a periglacial environment

Growing pressures on major wetlands and efforts to protect them, such as the Ramsar Convention

The Ramsar Convention—the Convention on Wetlands—was signed in Ramsar in Iran in 1971. It is a treaty that was signed by a number of governments in order to protect the unique ecology of wetlands. The UN Environment Programme (UNEP) estimates that about 6% of the world's land surface can be classified as wetlands.

Case study: The future possibilities for one wetland area—the Iraqi Marshes

The Iraqi Marshes are an area of wetlands in the south-east of Iraq. The area is home to the Marsh Arabs, who live on the water in a way that is unique to this part of the Middle East. In recent decades, their life has been made difficult due to Saddam Hussein draining the marshes to punish the Arabs for being disloyal during the Iran–Iraq war in the 1980s and the invasion of nearby Kuwait in 1990. By draining water from the marshes via a series of dams upstream, Hussein forced the Marsh Arabs out of the area, displacing over 200,000 people. By 2003, the wetlands had shrunk by 90%. Livestock perished, revenue decreased, and the unique culture of the Marsh Arabs was devastated.

The UN began a programme to restore the marshes in 2003 and by 2006, 50% of the marshes had been rejuvenated.

In 2016, the area was designated as a UNESCO World Heritage site, and by 2017 some families had returned to the area thanks to further initiatives from the Iraqi government.

Several thousand families now make a living from fishing and buffalo herding, while native birds are present once more. Domestic tourism has been increasing, with groups touring the wetlands and watching the migratory birds such as eagles and pelicans. However, the situation with sectarian conflict in Iraq and the rise of ISIS has meant that few tourists from outside Iraq have visited.

There are positive possibilities, but there are some negative issues too. Turkey is planning to build a series of dams that will starve the rivers Euphrates and Tigris of potentially 40–50% of their normal waters and thus once again decimate the Iraqi wetlands.

Test yourself

A.21 Define the term "wetlands".

A.22 Outline the value of wetlands.

QUESTION PRACTICE

The graph below shows the number of floods per decade for a river.

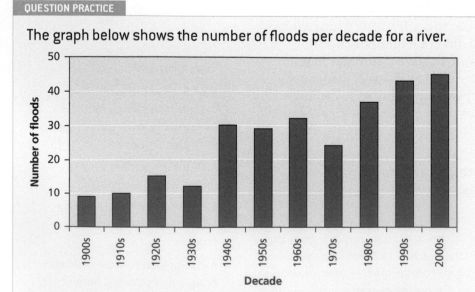

Source of data: International Baccalaureate Organization (2016)

a) Describe the changes in flood frequency shown on the graph. [2]

b) Outline one flood prediction strategy. [2]

c) Suggest one physical reason and one human reason why the risk of a river flooding can change over time. [3 + 3]

Essays

Either: Examine the role of river deposition in the formation of floodplain landforms. [10]

Or: Evaluate the costs and benefits of river flooding. [10]

How do I approach these questions?

a) This 2-mark question will require two distinct points and the inclusion of data. Try to categorize changes that have taken place over time and write one sentence for each time period. Consider the general trend, periods of significant increase or decrease and any anomaly that is not part of the general trend. Do not forget to include data in your answer.

b) A brief account or summary is required for this question; it is worth 2 marks so two or three sentences will be sufficient.

c) For each factor (one natural and one artificial) it will be necessary to clearly state the factor, and then explain in detail how it links to increased or reduced flooding. It would also be appropriate to try and include an example for each of the factors.

First essay choice:

The command term is "examine", and because of this you will need to ensure that the relationships between the river processes and the formation of landforms on a floodplain are discussed. Consider how some landforms are a product of deposition and other landforms require other fluvial processes. Try to ensure that your paragraphs have a focus, and be prepared to draw an annotated sketch that explains how **processes** help to form different landforms on the floodplain.

Second essay choice:

"Evaluate" is the command term for the alternative essay response, and therefore you will need to ensure that you offer a measured response, with both costs and benefits detailed in your answer. In order to provide context and to substantiate the positive and negative effects, you will need to include relevant case studies with detail. Conceptual connections are possible: you can outline how flooding can be beneficial to a **place** in terms of how the land around a river is used.

SAMPLE STUDENT ANSWER

a) The general trend of flood frequency from the 1900s to the 2000s is increasing. Between the 1900s and 1920s is quite a large jump from about 9 floods to approximately 15 floods. There is a decrease between 1920s and 1930s from approx. 15 floods to about 11 floods in the 1930s. Between 1930s and 1940s is the largest increase from 11 floods to 30 floods in 1940s. Relatively staying the same between 1940s to 1960s but a small decrease in 1950s.

▲ General increase a valid point

▲ Significant increase a valid point

This is a comprehensive answer that includes plenty of data and achieves full marks. Given the time constraints in an exam, it would be more appropriate to include less content since two sentences would have been sufficient.

Increasing number of floods = 1 mark

Flood-rich period 1940s–1960s = 1 mark

Marks 2/2

15

b) Weather forecasting and tracking weather systems is one way that authorities are able to predict floods.

This answer includes a valid strategy, but it requires development in order to link the strategy to flood prediction. For example, the monitoring of any future increases in precipitation can aid authorities in making predictions for potential flooding when also considering river levels and soil moisture. The answer only states or identifies a strategy rather than providing an outline.

Mark 1/2

▲ Increased precipitation

▼ Lacking development/
explanation/examples

▲ Channelization

c) The amount of precipitation received in an area can affect the discharge of a river as additional water will fall in the channel. The human interferences such as flood management, for example channelization, can create negative effects. This is because places further downstream that have not implemented channelization will see an increase in discharge resulting in flood as the size of the channel cannot cope with the excess water.

Increased precipitation, but quite simplistic development = 1 mark

Channelization = 1 mark

Explanation is present via a reduction in capacity for places downstream to cope with an increase in discharge = 1 mark

This answer requires more depth. For the physical reason, precipitation could be further developed by explaining that additional volume will be added to the river via overland flow and throughflow. In addition, antecedent moisture could be present from a previous rainfall event, which would further increase overland flow and the risk of flooding. An example would ensure that 3 marks for the physical factor would be credited. The human part of the answer is better since there is more explanation, and an example of where channelization has caused flooding would provide the third mark for this part of the answer.

Marks 3/6
Essay

Either: Examine the role of river deposition in the formation of floodplain landforms.

Floodplain landforms are caused by different river processes the further down or upstream a landform is.

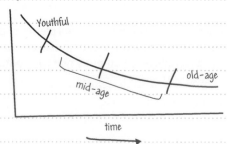

Due to discharge, and velocity upstream and gradient also playing a major role, the landforms may be different due to erosion taking place such as hydraulic action, abrasion and attrition.

▲ Processes

Eroding the large rocks and turning to matter that can be transported downstream and the formation of larger levees begin to form as the velocity of the rivers gets slower the centre of the river will be the fastest as it will be the largest wetted perimeter leaving less friction hence deposition on the sides of the river as less velocity and matter cannot be transported as it's too heavy therefore it is deposited.

▲ Landform

▼ This is a very long sentence and it should be broken down in order to achieve greater clarity

▲ Deposition—link to the question

▲ Processes

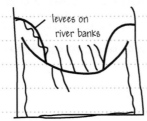
levees on river banks

Just like the formation of oxbow lakes the velocity is higher on the inner side of the river making it deposit on the outer.

▲ Landform

▼ This is incorrect; the velocity is higher on the outer side of the meander, and will deposit on the inner side.

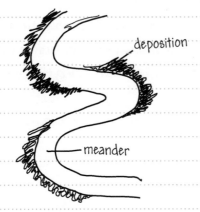
deposition

meander

Rivers have different stages such as the upper, the middle and the lower courses and these sections will have different landforms present as the influence of erosion and deposition will differ.

Some sentences are very long, and the essay would benefit from being more structured with clear, focused paragraphs. Some appropriate terminology has been included although more could have been included such as 'slip-off slopes' for the inside of a 'meander'. The drawings do not add much value to the response since they lack annotations and a limited amount of content is provided in the response. Floodplains and deltas could have been discussed, as well as providing a clearer explanation of the formation of a meander.

Marks 4/10

Or: Evaluate the costs and benefits of river flooding.

▲ Appropriate terminology

River flooding tends to occur on an annual basis in most basins due to climatic conditions such as snowmelt in spring, monsoonal rains, and more. The effects of this flooding can be detrimental. However, the relative consistency of these means that communities living within floodplains can adequately prepare in order to avoid detrimental effects and even benefit from flooding. An example of this is Bangladesh, where 30% of the country floods annually due to a combination of its three main rivers – the Brahmaputra, Ganges and Meghna – flooding

▲ Relevant example with detail

due to Himalayan snowmelt in the spring paired with heavy monsoonal rains. These floods lead to improved soil fertility

▲ Appropriate terminology

(as sediment/alluvium is deposited over the floodplain), which

▲ Benefit

benefits agricultural productivity. This also causes the country's

▲ Benefit

aquifers to be restored, increasing the groundwater supply. This is highly important in a highly densely populated country like Bangladesh, which may be at risk of water scarcity without these

▲ Application to example

floods. Due to the inexistence of sewage treatment plants due to the country's poverty, water cannot always be extracted from the river, and therefore the natural filtration system of aquifers is

▲ Application to example

even more important.

Floods also help to flush away these pollutants, increasing the overall well-being of local ecosystems as well as the aforementioned benefit of eradicating the country's high

▲ Benefit

pollution in its rivers.

As can be seen, the effects of these annual floods can be

highly beneficial, especially for a nation like Bangladesh that doesn't have very developed urban settlements and relies primarily on agriculture. However, the volatility of global climate on an increasing scale due to climate change means that these floods are not entirely predictable, and can have devastating negative consequences.

Normal floods can destroy settlements, kill livestock and destroy infrastructure, however their unpredictability means their consequences can be mitigated. Intense floods, such as the Bangladesh flood of 2017, can be devastating due to their unexpected nature. This particular case was caused by synchronized discharges of the three rivers due to abnormally intense monsoon rains at the same time as high snowmelt. The effects were consolidated by Himalayan foothill deforestation as well as a high water table due to consistent antecedent rainfall. This led to the largest flood in modern history and devastated the low-lying delta nation. It killed over 100 people and 6 million people were affected. Its effects were consolidated by Bangladesh's poverty, however the flood's negative consequences were immense. Here it can be seen that despite consistent flooding being beneficial, the potential for a large-scale unexpected flood can arise and be highly damaging.

In conclusion, I disagree with the statement that the negative consequences always outweigh benefits, given that it is often the benefits of flooding that maintain the survival of these river/delta communities. However, climatic volatility can lead to large floods which communities aren't prepared for, which can often be devastating enough to make the benefits of flooding appear meagre in comparison.

- ▲ Case study knowledge
- ▼ This final sentence is not aligned with the rest of the paragraph, which discusses the benefits of flooding
- ▲ Structure
- ▲ Negative consequences
- ▲ Relevant example
- ▲ Appropriate terminology
- ▲ Appropriate terminology
- ▲ Negative impacts, case study detail
- ▲ Effective summary with no new information included, aligned with the rest of the essay

Detailed explanation of the problems and benefits with an evaluation.

Marks 10/10

B OCEANS AND COASTAL MARGINS

Covering more than 70% of the Earth's surface, oceans are of great importance to humans in a number of ways. This optional theme provides an introduction to the physical characteristics and processes of the oceans with particular reference to the atmosphere–ocean linkage, concentrating on the important role that oceans play in influencing climatic conditions.

You should be able to show:

✓ how physical **processes** link the Earth's atmospheric and ocean systems;

✓ how coastal **places** are shaped by their interactions with oceans;

✓ the varying **power** of different stakeholders in relation to coastal margin management;

✓ future **possibilities** for managing the oceans as a global commons.

B.1 OCEAN–ATMOSPHERE INTERACTIONS

• **ENSO (El Niño Southern Oscillation)** – a reversal of the normal atmospheric circulation in the southern Pacific Ocean.

• **La Niña** – an intensification of normal atmospheric processes in the southern Pacific Ocean.

You should be able to show how physical processes link the Earth's atmospheric and ocean systems:

✓ The operation of ocean currents, including their distribution, nutrient and energy transfers, and the importance of oceanic conveyor belts;

✓ Atmosphere–oceanic interactions associated with El Niño Southern Oscillation (ENSO) and La Niña cycles and their climatic, environmental and economic effects;

  ✔ Detailed examples of the geographic impacts of El Niño and La Niña;

✓ The formation, distribution and physical impacts of hurricanes on coastal margins, including storm surges;

  ✔ Case study of one hurricane and its impacts on coastal places and people;

✓ The changing role of oceans as a store and source of carbon dioxide (CO_2) and the impacts of ocean acidification on coral reefs.

The operation of ocean currents

Surface ocean currents are caused by the influence of prevailing winds blowing steadily across the sea.

The effect of ocean currents on temperatures depends on whether the current is cold or warm. Warm currents from equatorial regions raise the temperatures of polar areas (with help from prevailing westerly winds).

Ocean currents are important as they transfer heat, oxygen and nutrients around the world.

Nutrient and energy transfers

Many eastern oceans experience upwelling currents, where the ocean currents move cold water, rich in nutrients, from the ocean floor to the surface. Such upwelling currents are found off the coast of Peru, California and south-west Africa.

Test yourself

B.1 Identify (a) one warm ocean current and **(b)** one cold current in the North Atlantic Ocean. [2]

B.2 Outline the impact of ocean currents on the climate of places at coastal margins. [4]

The importance of oceanic conveyor belts

In addition to the transfer of energy by wind and ocean currents, there is also a transfer of energy by deep sea currents. Oceanic convection occurs from polar regions where cold salty water sinks into the depths and makes its way towards the equator. The densest water is found in the Antarctic area. This cold dense water sweeps round Antarctica at a depth of about 4 km. It then spreads into the deep basins of the Atlantic, the Pacific and the Indian Ocean. Surface currents bring warm water to the North Atlantic from the Indian and Pacific Oceans. These waters give up their heat to cold winds which blow from Canada across the North Atlantic. This water then sinks and starts the reverse convection of the deep ocean current.

Temperature, salinity and pressure affect the density of seawater. Large water masses of different densities are important in the layering of the ocean water (denser water sinks). A cold, highly saline, deep mass of water is very dense, whereas a warm, less saline, surface water mass is less dense. When large water masses with different densities meet, the denser water mass slips under the less dense mass.

Atmosphere–oceanic interactions associated with ENSO and La Niña cycles

El Niño conditions in the Pacific Ocean

El Niño—the "Christ Child"—is a warming of the eastern Pacific that occurs at intervals between two and ten years, and lasts for up to two years.

During El Niño Southern Oscillation (ENSO) episodes, water temperatures in the eastern Pacific rise as warm water from the western Pacific flows into the east Pacific. The sea surface temperatures (SSTs) of over 28°C extend much further across the Pacific. Low pressure develops over the eastern Pacific; high pressure over the west.

The 2015–16 El Niño event led to 100 million people being short of water and food. Some 40 million people in rural areas and nearly 10 million people in urban areas in southern Africa were affected. The area experienced its driest year for 35 years. Up to one million children in East Africa and southern Africa needed treatment for malnutrition. The heavy rains in South America enabled the Zika virus to spread easily (there was more stagnant water for mosquitoes to breed).

La Niña

La Niña is an intermittent cold current that flows from the east across the equatorial Pacific Ocean. It is an intensification of normal conditions whereby strong easterly winds push cold upwelling water off the coast of South America into the western Pacific. Its impact extends beyond the Pacific and has been linked with unusual rainfall patterns in the Sahel and in India, and with unusual temperature patterns in Canada. Generally, there are wetter-than-normal conditions over northern Australia, Indonesia and the Philippines. Wetter conditions are also experienced in northern Brazil and south-eastern Africa. The summer monsoon tends to be stronger over north-west India. Drier conditions are experienced along the west coast of South America.

> **Concept link** 🔗
>
> **PROCESSES:** The processes that exist within the hydrosphere and atmosphere, and between both systems, enable the distribution of energy. These natural processes are important at a variety of scales, such as the economic impact on a country's agricultural industry in the South Pacific during a La Niña year, for example. Human input can alter some of these processes, such as the increased frequency of cyclones due to the warming of the atmosphere and oceans.

Test yourself

B.3 Compare the climate conditions in the southern Pacific region associated with El Niño events with those of La Niña. [2+2]

The formation, distribution and physical impacts of hurricanes on coastal margins

Hurricanes are intense hazards that bring heavy rainfall, strong winds and high waves; they cause other hazards such as flooding and mudslides. High-intensity rainfall, as well as large totals (up to 500 mm in 24 hours), invariably causes flooding.

Hurricanes move excess heat from low latitudes to higher latitudes.

For hurricanes to form, a number of conditions are needed:

• Sea temperatures must be over 27°C. (Warm water gives off large quantities of heat when it is condensed—this heat drives the hurricane.)

• The low pressure area has to be far enough away from the equator so that the Coriolis force (the force caused by the rotation of the Earth) creates rotation in the rising air mass. If it is too close to the equator, there is insufficient rotation and a hurricane will not develop.

• There must be low wind shear (the gradient of wind velocity with altitude) to allow the hurricane to maintain its structure.

Case study: Hurricane Sandy, 2012

In Haiti, over 100 deaths were attributed to Sandy. Tents across the refugee camps were flooded and crops washed away.

In the USA, Sandy affected 24 states, including the entire eastern seaboard from Florida to Maine and west across to Michigan and Wisconsin. Damage was particularly severe in New Jersey and New York. At least 131 people in eight states were killed. The New York Stock Exchange was closed for two days, the first weather closure since 1985 and the first two-day weather closure since 1888. Seven subway tunnels under the East river were flooded. Battery Park in Manhattan had a storm surge of 14 feet. Over 100,000 homes on Long Island were damaged or destroyed.

The changing role of oceans as a store and source of carbon dioxide (CO_2), and the impacts of ocean acidification on coral reefs

The oceans are the largest reservoir of carbon in the carbon cycle—they contain approximately 38,000 trillion kg of carbon. The oceans are also a major *source* of carbon. For example, photosynthesis by plankton creates organic compounds from CO_2. Some of this passes through the food chain and sinks to the seabed where it is decomposed into sediments.

Acidification of oceans

Atmospheric carbon dioxide is taken in by the ocean and becomes dissolved. It reacts with sea water to produce carbonic acid, lowering the water's pH level and making it more acidic. This affects coral reefs and organisms with shells.

Test yourself

B.4 Briefly **explain** the main hazards associated with hurricanes.   [4]

B.5 Suggest why Haiti experienced a relatively high number of deaths despite not being in the direct path of Hurricane Sandy.   [2]

Test yourself

B.6 Analyse the importance of the ocean as a source and a store of carbon.   [2+2]

B.7 (a) Describe how the ocean is becoming more acidic.   [2]

    (b) Explain how more acidic oceans are damaging to marine life.   [3]

B.2 INTERACTIONS BETWEEN OCEANS AND COASTAL PLACES

You should be able to show how coastal places are shaped by their interactions with oceans:

✓ Physical influences on coastal landscapes, including waves, tides, sediment supply, lithology, vegetation, subaerial processes and wave processes (littoral drift, hydraulic action and abrasion);

✓ The characteristics and formation of coastal landforms of erosion and deposition, including wave-cut platforms, cliffs, stacks, spits and beaches;

✓ Advancing and retreating coastlines, including the role of isostatic and eustatic processes, and the associated landforms (relict cliff, raised beach, fjord);

✓ The role of coastal processes, wind and vegetation in sand dune development.

Physical influences on coastal landscapes

Waves

Constructive waves tend to occur when wave frequency is low (6–8/minute). Swash is stronger than backwash, so sediment is moved up the beach.

Destructive waves are the result of locally generated winds, which create waves of high frequency (12–14/minute). As the backwash is stronger than the swash, material is eroded from the beach.

Tides

Tides are regular movements in the sea's surface, caused by the gravitational pull of the Moon and Sun on the oceans. In general:

• tides are greatest in bays and along funnel-shaped coastlines

• during low pressure systems, water levels are raised 10 cm for every decrease of 10 mb.

Sediment supply

Sediment transport is generally categorized into two modes:

• **Bedload**—grains transported by bedload are moved with continuous contact (traction or dragging) or by discontinuous contact (saltation) with the seafloor.

• **Suspended load**—grains are carried by turbulent flow and generally are held up by the water.

Subaerial and wave processes

Subaerial, or cliff-face, processes include salt weathering, freeze-thaw weathering, solution weathering, slaking and biological weathering.

Mass movements are also important in coastal areas, especially slumping and rock falls.

• **Eustatic changes** – worldwide changes in sea level caused by the growth and decay of ice caps, thereby locking up and later releasing water from ice.

• **Isostatic changes** – localized changes in the relative level of the land and sea, caused by the depression of the Earth's crust, such as due to the weight of an ice sheet. Following deglaciation, the crust beneath the weight begins to rise again, and relative sea level therefore falls.

• **Advancing coasts** – coastlines that are growing/getting larger either due to deposition or a fall in sea level.

• **Subaerial** – processes that occur on the Earth's surface.

» Assessment tip

Oceans and their coastal margins are open systems as they receive energy and matter from external sources. However, for convenience, geographers use the concept of a sediment cell (or littoral cell) in which inputs and outputs are balanced within a single bay or region, and each sediment cell is self-contained.

 Content link

Some of these mass movements are discussed in more detail in option D.1.

Test yourself

B.8 Distinguish between constructive and destructive waves. [2+2]

B.9 Describe what is meant by a littoral cell system. [2]

B.10 Explain the importance of the tidal range. [3]

B.11 Analyse the changes that occur as a result of wave refraction. [3]

Wave erosion

The processes of erosion are abrasion, attrition, solution (corrosion) and hydraulic action.

Littoral drift

Littoral (longshore) drift leads to a gradual movement of sediment along the shore, as the swash moves in the direction of the prevailing wind, whereas the backwash moves straight down the beach, following the steepest gradient.

The characteristics and formation of coastal landforms of erosion and deposition

Erosional landforms

Cliff profiles are very variable and depend on a number of controlling factors. One major factor is the influence of lithology (rock type), that is to say bedding (horizontal strata of sedimentary rocks) and jointing (cracks along vertical lines of weakness).

Wave-cut platforms are most frequently found in high-energy environments and are typically less than 500 metres wide with an angle of about 1°. Cliff- and shore-platform evolution means that steep cliffs are replaced by a lengthening platform and lower angle cliffs that are subjected to subaerial processes rather than marine forces.

Test yourself

B.12 Outline the difference between a stack and a stump. [2]

B.13 Analyse the formation of wave-cut platforms. [3]

Wave refraction concentrates wave energy on the flanks of headlands. If there are lines of weakness, these may be eroded to form a geo (a widened crack). Geos may be eroded and enlarged to form caves, and if the caves on either side of a headland merge, an arch is formed. Further erosion and weathering of the arch may cause the roof of the arch to collapse, leaving an upstanding stack. The eventual erosion of the stack produces a stump.

Depositional landforms

A beach is a feature of coastal deposition, consisting of pebbles on exposed coasts or sand on sheltered coasts.

Spits

These localized depositional features will develop where:

- abundant material is available, particularly shingle and sand

- deposition is increased by the presence of vegetation (reducing wave velocity and energy).

Test yourself

B.14 Describe the formation of spits. [3]

B.15 Briefly explain the formation of caves and arches. [2+2]

Spits are common along indented coastlines, for example, near Swapkopmund, Namibia. The long, narrow ridges of sand and shingle that form spits are always joined at one end to the mainland.

Advancing and retreating coastlines

Sea levels change in conjunction with the growth and decay of ice sheets. Eustatic change refers to a global change in sea level. The change in the level of the land relative to the level of the sea is known as isostatic adjustment or isostasy.

A simple sequence of sea-level change can be described:

1. Temperatures decrease, glaciers and ice sheets advance and sea levels fall, eustatically.

2. Ice thickness increases and the land is lowered isostatically.

3. Temperatures rise, ice melts and sea levels rise eustatically.

4. Continued melting releases pressure on the land and the land rises isostatically.

Features of emerged coastlines include:

- raised beaches, such as along the west coast of Malta

- coastal plains

- relict cliffs, such as those along the Fall Line in eastern USA

- raised mudflats, for example, the mudflats on the south coast of the Rio la Plata, Argentina.

Submerged coastlines include:

- rias, such as the Georges River, Sydney, Australia—drowned river valleys caused by rising sea levels or due to a sinking of the land

- fjords, such as the Oslo Fjord—glacial troughs drowned by the sea

- fjards or "drowned glacial lowlands", for example, Somes Sound, Maine, USA.

The role of coastal processes, wind and vegetation in sand dune development

Sand dunes form where there is a reliable supply of sand, strong onshore winds, a large tidal range and vegetation to trap the sand. Strong winds transport a large volume of sand onshore, especially at low tide. Vegetation causes a reduction in wind velocity, especially in the lowest few centimetres above the ground, and this reduces energy and increases the deposition of sand.

The greater the amount of vegetation, the greater the amount of deposition, and more rapid sand dune development occurs.

Concept link

PLACES: Similar to other parts of the planet, coastal margins are dynamic places that can feature a range of different landforms that are subject to processes of erosion and deposition, both in their formation but also in their evolution. Thus places have quite unique characteristics. For example, coastlines with striking and impressive cliff and dune formations. Spatial change on these margins relies on global processes; a cliff may be eroded due to an increase in storm surges in oceans as global temperatures and acidification increase.

Test yourself

B.16 Describe the processes that form (a) fjords and (b) relict cliffs.   [2+2]

Test yourself

B.17 Analyse the formation of sand dunes.   [3]

B.3 MANAGING COASTAL MARGINS

You should be able to show the varying power of different stakeholders in relation to coastal margin management:

✔ Coastal erosion and flooding management strategies, including cliff-line stabilization and managed retreat;

   ✔ One coastal management case study focused on the decision-making process and perspectives of different actors;

✔ Conflicting land-use pressures on coastlines, including commercial land uses (tourism, industry and housing) and conservation measures;

   ✔ One case study to illustrate the roles of, and outcomes for, coastal stakeholders;

✔ Management of coral reefs and mangrove swamps, including different stakeholder perspectives on their use and value;

   ✔ Detailed examples of both ecosystems and their issues;

✔ Sovereignty rights of nations in relation to territorial limits along coastal margins and exclusive economic zones (EEZs).

- **Exclusive economic zone (EEZ)** — an area in which a coastal nation has sovereign rights over all the economic resources of the sea, seabed and subsoil, extending up to 200 nautical miles from the coast (one nautical mile is c.1.85 km).

- **Sovereign** — having independent authority over a territory.

Concept link

POWER: The value of coastal margins should not be understated, and different stakeholders view this value through different lenses. For example, commercial developers are interested in tourism revenue, while ecologists see coastlines as areas containing unique ecosystems that support biodiversity and protect against environmental degradation. The value of places can often lead to conflict and friction due to the contrasting motivations of parties who have a vested interest. These conflicts can exist at a local scale, but they have also taken place at an international scale in the past and present and could quite possibly take place in the future as countries seek to establish their territory in oceans.

Test yourself

B.18 Describe the advantages of **(a)** gabions and **(b)** cliff regrading.   [2+2]

B.19 Outline why some stakeholders may hold different viewpoints regarding coastal management schemes.   [6]

Coastal erosion and flooding management strategies, including cliff-line stabilization and managed retreat

Coastal hard engineering management strategies

Cliff-base management strategies include:

- Sea walls, which are large-scale concrete curved walls designed to reflect wave energy. They are easily made and good in areas of high density. However, they are expensive.

- Gabions, which are made from rocks contained in wire cages and absorb wave energy. They are cheaper than sea walls and revetments but only work on a small scale.

Cliff-face management includes:

- Cliff drainage, which removes water from rocks in the cliff.

- Cliff grading, which refers to the lowering of the cliff angle to make the cliff safer.

Coastal soft engineering management strategies

Soft engineering management strategies include:

- Beach nourishment, which uses sand pumped from elsewhere to replace the eroded sand.

- Managed retreat, which allows the coastline to be eroded in certain places.

Case study: Coastal management at Miami Beach, USA

By the 1950s there was little left of Miami Beach, USA, as groynes, sea walls and dredging had led to the removal of vast amounts of sand. Due to the importance of tourism and recreation, the beach was replenished and protected. During the 1970s and 1980s the US Army Corps of Engineers built a new beach, with 18 million cubic metres of sand. Around 0.25 million cubic metres are needed each year to replenish the sand eroded from the beach.

Conflicting land-use pressures on coastlines

There are a number of conflicting pressures on coastlines including commercial land use (tourism, industry and housing) and conservation measures. Tourism, housing developments and industry can lead to land-use changes (for example, building of houses, industries, hotels and tourist infrastructure). It can cause loss of habitats and species diversity, visual intrusion, lowering of groundwater tables, saltwater intrusion in aquifers and water pollution. In contrast, most conservationists want to preserve natural habitats and to limit the amount of damage that new developments cause.

Case study: The Red Sea reefs

Tourism is a growth industry in the northern part of the Red Sea. Resorts such as Hurghada and Sharm-el-Sheikh have attracted large numbers of visitors. However, some tourist developments may impact the attractions that people want to visit. Construction creates dust, which gets blown onto the reefs. Construction also creates new land from landfill which causes major disruption to reefs. Disposal of sewage is a problem, and even where it has been treated, algal growth on the reefs has occurred.

Marine parks have been created in the area (for example, Ras Mohamed Marine Park at Sharm-el-Sheikh) to aid reef conservation, but this has angered fishermen. There are many threats to the reefs, including tourism developments (sedimentation, habitat degradation, discharges), illegal fishing and mass tourism.

Management of coral reefs and mangrove swamps

Coral reefs are often described as the "rainforests of the sea" on account of their rich biodiversity. Coral reefs contain nearly a million species of plants and animals, and about 25% of the world's sea fish breed, grow, spawn and evade predators in coral reefs.

The value of coral reefs

Coral reefs are of major biological and economic importance.

There are many stakeholders with an interest in coral reefs, including fishermen, tourists and people involved in the tourism sector, conservationists and industrialists.

Threats to coral reefs

Global warming, sea-level rise, overfishing, destruction of the coastal habitat, as well as pollution from industry, farms and households are all endangering coral reefs.

Global climate change will cause irreparable damage to coral reefs in our lifetime for several reasons:

- Increasing sea surface temperatures will cause more coral bleaching.

- The abundance of many coral species will be reduced and some species may become extinct.

- Increasing ocean acidification will reduce calcification in corals and other calcifying organisms, resulting in slower growth, weaker skeletons and eventual death.

▲ Figure B.3.1. A coral reef in Antigua

Coral reefs: Possible management strategies

To avoid permanent damage and support people in the tropics, it is recommended that:

- emissions of greenhouse gases are reduced

- damaging human activities (sedimentation, overfishing, blasting coral) is limited

- more coral reefs are designated as Marine Protected Areas (MPAs) to act as reservoirs of biodiversity.

Mangrove swamps: Threats

Mangroves are salt-tolerant forests of trees and shrubs that grow in the tidal estuaries and coastal zones of tropical areas. Mangroves cover about 25% of the tropical coastline.

Mangroves provide humans with many ecological services. These include products such as fuelwood, charcoal, timber, thatching materials, dyes, poisons, as well as food such as shellfish and crustaceans. Many fish species, both commercially farmed and farmed for subsistence, use mangrove swamps and sea-grass beds as nurseries. In addition, mangrove trees provide protection from tropical storms, and act as sediment traps.

Test yourself

B.20 Analyse the conditions necessary for the growth of coral.   [4]

B.21 Using examples, **analyse** the variety of pressures that affect coral reefs.   [4]

B.22 Suggest the ecological services provided by mangrove swamps.   [3]

Test yourself

B.23 Distinguish between territorial waters and exclusive economic zones.   [2]

B.24 Explain why conflicts arise over the use of EEZs.   [4]

Owing to the large range of benefits that mangroves provide, many stakeholders are interested in mangrove swamps, including fishermen, farmers, conservationists, local residents and politicians.

Mangrove swamps: Possible management strategies

Management strategies include:

- restoration and afforestation
- managed realignment—allowing mangroves to migrate inland
- generic protection of mangrove ecosystems.

Sovereignty rights of nations in relation to territorial limits along coastal margins and exclusive economic zones (EEZs)

Coastal states are free to exploit, develop and manage all resources found in the waters, on the ocean floor area extending 200 nautical miles from their shore. Territorial waters are the waters (sea/ocean) over which a country has full sovereignty rights. In contrast, the exclusive economic zone is the sub-surface area over which a country has exclusive rights for the exploitation of marine resources, for example, fish, energy resources and metals.

B.4 OCEAN MANAGEMENT FUTURES

- **Abiotic resources** – non-living resources such as oil and gas.
- **Biotic resources** – living resources such as fish and vegetation.
- **Geopolitical** – the influence of geographic factors (location, resources) over politics and power.

Concept link

POSSIBILITIES: With growing pressures placed on ocean environments, there is an increased need to ensure that initiatives are implemented and managed to achieve sustainability. The Sustainable Development Goals contains targets that represent positive possibilities from strategies applied throughout the world. However, the management of oceans is a complex matter; the resulting possibilities can also be negative, especially when trying to cope with growing consumerism and increasing geopolitical tension, and thus power and scale cannot be ignored.

You should be able to show examples of future possibilities for managing the oceans as a global commons:

- Causes and consequences of increasing demand for the abiotic resources of oceans, including minerals, oil and gas;
- Trends in biotic resource use (fish and mammals) and the viability of alternatives to overfishing, including aquaculture, conservation areas and quotas;
- Strengths and weaknesses of initiatives to manage ocean pollution, including local and global strategies for radioactive materials, oil and plastic waste;
- The strategic value of oceans and sources of international conflict/ insecurity, including the contested ownership and control of islands, canals and transit choke points;
  - One contemporary geopolitical case study focusing on a contested ocean area.

Causes and consequences of increasing demand for the abiotic resources of oceans

As the world's population grows, its economies develop and standards of living rise, the demand for raw materials increases, especially for non-renewable resources such as oil and gas.

The Arctic could hold a quarter of the world's undiscovered gas and oil reserves. This amounts to 90 billion barrels of oil and vast amounts of natural gas. Nearly 85% of these deposits are believed to be offshore. Canada, Denmark, Norway, Russia and the USA are racing to establish the limits of their territory, stretching far beyond their land borders. They are competing to gain better access to the Arctic's resource base.

Trends in biotic resource use (fish and mammals) and the viability of alternatives to overfishing

World fisheries and aquaculture contributed almost 171 million tonnes of fish in 2016, valued at over US$362 billion. The world's supply of fish as food has grown dramatically since 1961, with an average growth rate of 3.2% per year compared with a growth rate of 1.6% per year for the world's population.

Overfishing

Nearly 70% of the world's stocks are in need of management. Cod stocks in the North Sea are less than 10% of 1970 levels. Fishing boats from the EU regularly fish in other parts of the world.

Aquaculture

Aquaculture involves raising fish commercially, usually for food. Between 2000 and 2016, world aquaculture grew by, on average, 5.8% per year. World aquaculture production in 2016 was estimated at 80 million tonnes.

Quotas and conservation areas

In the past, quotas, bans and conservation areas have failed to address the real problem of the fishing industry: too many fishermen chasing too few fish and too many young fish. For fisheries to be protected, the number of boats and fishermen need to be reduced.

Strengths and weaknesses of initiatives to manage ocean pollution

The strengths of initiatives to manage ocean pollution include increased public awareness, more legislation, and the work of civil society organizations, such as Greenpeace. However, there are weaknesses, including the size of the areas to be managed, the increasing number of pollutants (for example, plastic) and the origin of oceanic pollutants, many of which are land based. Marine-based pollution is also important—just under 50% of the great Pacific garbage patch derives from the plastic in fishing nets.

The strategic value of oceans and sources of international conflict/insecurity

A number of oceans have strategic importance. This may be due to the importance of shipping routes (for example, the Straits of Malacca, a transit choke point), control of islands (for example, the Falkland Islands), the presence of biotic and abiotic resources (for example, the Arctic Ocean), or the importance of canals for transport (for example, the Panama Canal).

Case study: Geopolitical conflict in the South China Sea

Countries with borders on the South China Sea include China, Taiwan, the Philippines, Malaysia, Brunei, Indonesia, Singapore and Vietnam.

The South China Sea has a very important strategic value. It is the second most-used sea lane in the world. Over 1.6 million cubic metres (10 million barrels) of crude oil a day are shipped through the Straits of Malacca. Moreover, the region has proven oil and gas reserves. Natural

Test yourself

B.25 Define the term "abiotic resource". [1]

B.26 Identify two abiotic resources found in the Arctic. [1]

B.27 Suggest why the Arctic is important for abiotic resources. [2]

B.28 Explain why development of abiotic resources in the Arctic may be damaging to the environment. [3]

>> **Assessment tip**

When describing data, remember to refer to the maximum, minimum, trend and anomalies.

Test yourself

B.29 Analyse the reasons why two different seas/oceans have developed as pollution hotspots. [2+2]

B.30 Describe how the physical geography of oceans is related to oceanic pollution. [2]

Test yourself

B.31 Suggest reasons why there is competition among countries for rights over the South China Sea. [2]

gas reserves are estimated to total around 7,500 cubic kilometres (266 trillion cubic feet). The countries are competing in their claims to develop these resources and to control sections of the South China Sea.

QUESTION PRACTICE

The following graph shows the distribution of hurricanes.

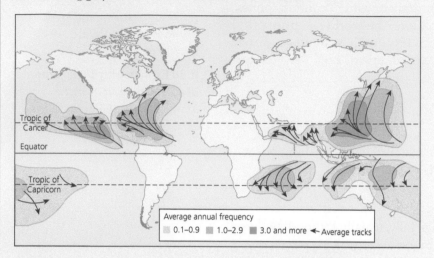

a) Describe the distribution of hurricanes as shown on the map. [2]

b) Explain two factors needed for the formation of hurricanes. [2+2]

c) Using examples, **explain two** reasons why oceans may become a source of geopolitical conflict. [2+2]

Essays

Either: Examine the advantages and disadvantages of different types of coastal management schemes. [10]

Or: Using a located example, **examine** how conflicting pressures in a coastal area have been managed. [10]

How do I approach these questions?

a) This question asks for a description. You should use the information on the map, for example, lines of latitude, compass bearing, major concentrations and some smaller ones.

b) You are asked to explain two factors. Each explanation is worth two marks, so some detail/development is needed for the award of the second mark.

c) Similarly, here you are asked to explain two reasons. Each explanation is worth two marks, so some detail/development is needed for the award of the second mark.

Essays

For these essays, the command term is examine. This means that you should discuss the underlying assumptions and interrelationships of the issues presented in the question. Good answers will relate these essay questions to the 4Ps (places, processes, power and possibilities) or other geographical concepts. Remember to refer to specific examples, and some critical thinking and evaluation should be present in your answer.

SAMPLE STUDENT ANSWER

▲ Valid points highlighted

a) The distribution of hurricanes is mainly in tropical areas, away from the equator, and mostly on the western side of oceans. There are some exceptions such as on the NE side of Australia and off the west coast of central America.

Marks 2/2

▲ Valid factor

▲ Development point

b) Sea temperatures must be over 27°C. (Warm water gives off large quantities of heat when it is condensed - this is the heat which drives the hurricane.)

The low pressure area has to be far enough away from the equator that the Coriolis force (the force caused by the rotation of the Earth) creates rotation in the rising air mass – if it is too close to the equator there is insufficient rotation and a hurricane would not develop.

▲ Valid second factor

▲ Development point

Marks 4/4

c) Some parts of oceans, such as the South China Sea in the Pacific, may become sources of geopolitical conflict between nations (e.g. China, Brunei, the Philippines, Vietnam and Indonesia) because they contain valuable resources such as oil, gas and fisheries. There can also be conflict over shipping channels e.g. the Straits of Malacca between Singapore and Indonesia.

▲ Valid point—oceans and competing nations

▲ Valid point—competition over resources

▲ A second conflict

▲ Development point

Marks 4/4

Essays

Either: Examine the advantages and disadvantages of different types of coastal management schemes.

There are many types of coastal management schemes. They are generally designed to reduce erosion and prevent flooding. Coastal management schemes vary in terms of their effectiveness and their cost.

Coastal management schemes are often divided into hard engineering and soft engineering schemes. Hard engineering schemes are those that physically alter the landscape, whereas soft engineering schemes try to work with nature. One of the simplest types of hard engineering schemes is a sea wall. This is a large-scale wall designed to reflect wave energy. Modern sea walls are often curved to redirect the waves away from the beach. Traditional sea walls were often straight and vertical and led to the scouring of the bed, and undermining of the walls' foundations. They are relatively expensive and last about thirty years. Chesil Beach, in Southern England, has curved sea walls protecting the village of Chiswell.

Another form of hard engineering are gabions. These are wire boxes which are filled with rocks. The wire boxes prevent the rocks from being washed away or used for abrasion, and the rocks reduce the energy of the wave as it passes between different rocks. They are cheaper than sea walls and, from a distance, can look attractive. The gabions on top of Chesil Beach (gabion mattresses) protect Chiswell from large shingle being flung in storm conditions.

▲ Simple introduction

▲ Good—distinction between major types of coastal management

▲ First example

▲ Good point—advantages of sea walls

▲ Good point—outlines a disadvantage

▲ Good point—develops disadvantage

▲ Good point—located example

▲ Clear account of sea walls with support

▲ Second type of management

▲ Description

▲ Function

▲ Advantage

▲ Located example

▲ Second method, with supporting example

▲ Third method

▲ Advantages and disadvantages

▲ Located examples

▲ Two more methods identified

▲ Advantage of cliff drainage

▲ Advantages of cliff regrading

▼ Clear advantage but no support

▼ Duplication of what's been said in introduction

▲ First type of soft engineering

▲ Good example

▲ Second type of soft engineering

▲ Advantages and disadvantages of managed retreat

▼ Interesting points here— could have been further developed

Groynes are wooden/concrete barriers placed at right angles to the shoreline to trap sediment being carried away by longshore drift. They can be successful at keeping sediment in an area but may lead to sediment starvation further down the coast, and lead to increased erosion down-drift. This occurred at Barton-on-Sea following the construction of groynes at Bournemouth and Boscombe.

Another form of coastal management is cliff-face strategies. These include cliff drainage and cliff regrading. Cliff drainage removes excess water that is contained within the rocks forming the cliff. This can help reduce the risk of mass movement, especially slumping. Cliff regrading reduces the slope angle of the cliff, and also makes slumping less likely.

"Soft engineering schemes" refers to methods that work with nature. This includes beach nourishment, which is adding sand or shingle to restore a beach. Miami Beach in Florida has been restored regularly. It looks attractive but is expensive and is only a short-term solution. In some areas, such as at Barton in southern England, there is managed retreat, in which some parts of the coastline are sacrificed whereas others are protected. At Barton, houses and businesses are protected but the camping and caravan site is allowed to be eroded – mobile homes are moved back, further away from the shoreline. This is quite unpopular, but it is cost-effective and allows nature to takes its course.

Coastal management schemes are costly. Also, the nature of the threat is changing with global climate change and rising sea levels (and in some cases sinking land). Schemes that may have worked years ago may well become obsolete as seas become more powerful in certain locations.

Valid account. Support generally present. The idea of cost could have been developed, and related to LICs.

Marks 8/10

Or: Using a located example, **examine** how conflicting pressures in a coastal area have been managed.

▲ Clear location

▼ Largely scene-setting/ background information

Studland Beach, on the south coast of Britain, is an excellent example of a coastal area where the needs of different recreational users and the environment are carefully managed. Studland Beach is still wild and largely unspoilt. It is has broad sandy beaches and a succession of sand dune ridges and slacks leading to heathland, woodland, marsh and a large fresh-water lake. Because of the rich and varied plant and animal life, the area was declared a National Nature Reserve in 1962.

The National Trust try to run Studland Beach as a family beach, and the numbers of visitors are controlled by price and car parking spaces available. Since the National Trust became responsible it has increased the capacity for car parking by the provision of an overflow car park that can take a maximum of 500 cars. There is also a daily litter collection. The Visitor Centre was built in 1990 at the Knoll Car Park which includes a shop, cafe and information point. There are 311 beach huts situated along the front of the beach and the majority of these are privately owned, but the Trust manages 38 for weekly and seasonal lets. There are several problems for the staff to deal with on the Beach.

There are many problems in the management of Studland Beach. There are conflicts between naturists and walkers along the coastal path, between power boat enthusiasts and sailing boaters, between board sailers, swimmers and sunbathers, and between dog walkers and visitors.

There are four car parks in Studland plus two overflow parks. The capacity on a busy day is 3,500 cars. There is also roadside parking along the Ferry Road of approximately 1,000 cars and 300 around the village on verges etc. During the season, Easter to end September, total number of cars in National Trust parks are between 140,000 and 200,000. On a busy day there are about 25,000 people on the beach and up to 1,000 cars parked on the Ferry Road. As many as 300 cars are parked illegally on yellow lines in the village.

There are more than 1,000,000 visitors per year. These are concentrated in time. Most arrive in July and August. The estimated numbers of visitors on the beach on a busy day is 20,000–25,000 with approximately 8,000 coming by foot across on the Bournemouth/Studland chain ferry. Of all the visitors about 95% come just for the beach and only 5% venture into the Nature Reserve, which contains the sand dunes, heathland, woodland and Little Sea Lake.

There is a large volume of litter – 12–13 tonnes a week are dumped on the beach due to the high influx of visitors. To cope with the problem the National Trust put out approximately 200 litter bins during the summer months. Another problem is lost children, often up to 30 a day, who have to be reunited with their parents.

Annotations:
- ▼ Introduces located example. Focus is on conflicting tourism/recreation users rather than conflicting pressures
- ▼ Good detail but needs to relate back to the question
- ▲ Some management identified
- ▲ Clear evidence—management of parking and litter
- ▼ Not yet related to conflict
- ▼ Needs to focus on conflicts
- ▲ Now focused on conflicts
- ▼ Identifies conflict between users but mainly implied
- ▼ Good detail but needs to focus on conflicting pressures
- ▼ Needs to focus on conflicting pressures
- ▼ Descriptive account of problems—could easily relate to conflict, e.g. between tourists and local residents over illegal parking
- ▼ Good detail but needs to focus on conflicting pressures
- ▼ Needs to refer back to the question
- ▼ Lots of very good information that could easily be related back to conflicting pressures
- ▼ This is about managing the impacts of tourism, not the conflicts

33

▲ Back on track!

▲ Management of conflicting pressures

▲ Finally! Conflict between walkers and naturists

▲ Good point—would be useful to define land-use zoning

▲ Describes land-use zoning—implied way of dealing with conflicting pressures

▲ That is to say, managing conflicting pressures

Dealing with conflicts of interest - the main one being the naturists. People using the coastal path for walking feel offended by the naturists. To this end a new footpath was created called Heather Walk which runs behind the naturist area in the dunes, which allows the general public to walk the full length of the beach without encountering naturists. There is some wear and tear on the footpaths on the Nature Reserve, but this is being monitored and walkways have been put down in the most worn parts.

The main way the beach has been managed is by land-use zoning. The car parks are at either end of the beach, and this is where the cafes and toilets are located, and these areas get most of the visitors. The nature reserve is located near the middle of the beach, and few people walk there. The naturist beach is also located around this area, so as to be as far from the family beaches as possible. Overall, it is a very effective way of managing the beach.

Lots of good detail, but lacks focus on conflict. Had the candidate said "To manage conflict between family tourism and nature conservation needs, land-use zoning has been used" this would have made the answer much more appropriate. Overall 6/10 as conflict and management are largely implicit.

Marks 6/10

C EXTREME ENVIRONMENTS

This option requires a study of **two** different extreme environments:

- Cold and high-altitude environments (polar, glacial areas, periglacial areas and high mountains in non-polar latitudes);

- Hot, arid environments (hot deserts and semi-arid areas).

Extreme environments are relatively inaccessible and are also difficult for human habitation. Nevertheless, there are possibilities for settlement and economic activity. This option examines the natural processes and the main landscape characteristics of both kinds of extreme environments, the way in which people have adapted to these environments and the opportunities they offer, and the challenges they pose for management and sustainability.

You should be able to show:

✔ why some **places** are considered to be extreme environments;

✔ how physical **processes** create unique landscapes in extreme environments;

✔ the varying **power** of different stakeholders to extract economic value from extreme environments;

✔ future **possibilities** for managing extreme environments and their communities.

C.1 THE CHARACTERISTICS OF EXTREME ENVIRONMENTS

You should be able to show why some places are considered to be extreme environments:

✔ Global-scale distribution of cold and high-altitude environments and hot, arid environments (polar, glacial areas, periglacial areas, high mountains in non-polar places) and hot, arid environments (hot deserts and semi-arid areas);

✔ Relief and climatic characteristics that make environments extreme, including unreliability and intensity of rainfall in arid environments and the risk of flash floods;

✔ How relief, climate, human discomfort, inaccessibility and remoteness present challenges for human habitation and resource development;

   ✔ Detailed examples for illustrative purposes;

✔ The changing distribution of extreme environments over time, including the advance and retreat of glaciers and natural desertification.

Global-scale distribution of cold and high-altitude environments and hot, arid environments

Cold arid environments

The world's main ice masses exist at high latitudes, as in Antarctica and Greenland, and at high altitudes, as in the Alps and the Himalayas.

> • **Arid** – having less than 250 mm of precipitation per year.
>
> • **Semi-arid** – having less than 500 mm of precipitation per year.
>
> • **Ice sheet** – a continent-size ice mass, at least 50,000 km^2 in size.
>
> • **Glacier** – a slow-moving body of ice.

Concept link

PLACES: Places that are classified as having extreme environments are those that present challenges for those attempting to live there and develop resources. Over time, the distribution of these areas is changing, due to processes such as global climate change, which can make some areas more habitable and other areas less. As a result of these challenges and opportunities, places evolve.

Approximately a quarter of the Earth's surface is characterized by long periods of extreme cold and winter snowfall, but it also has a summer thaw. These areas are known as "periglacial" areas and are widespread in the northern hemisphere including much of northern Canada and large parts of Siberia. There are also some high-altitude areas in low latitudes that experience periglacial conditions.

Hot, arid environments

Most arid areas are located in the tropics, associated with the subtropical high-pressure belt. However, some are located alongside cold ocean currents (such as the Namib and Atacama deserts), some are located in the lee of mountain ranges (such as the Gobi and Patagonian deserts), while others are located in continental interiors (such as the Sahara and the Australian deserts).

Relief and climatic characteristics that make environments extreme

Anchorage in Alaska, USA, is an example of a periglacial environment, whereas Casablanca in Morocco is an example of a semi-arid environment.

Anchorage has a mean annual temperature of −4°C. The temperature range is from 15°C in summer (July) to −11°C in winter (January), a range of 26°C. Rainfall is relatively low (374 mm), mainly falling between July and October. Much of the precipitation in winter will be in the form of snow.

In contrast, Casablanca's mean annual temperature is 18°C, ranging from a low of 13°C in winter to a high of 23°C in summer (July/August). This is a range of just 10°C. Rainfall is more plentiful in the winter months of October–February.

▲ **Figure C.1.1.** Oryx grazing in a hot, arid desert

How relief, climate, human discomfort, inaccessibility and remoteness present challenges for human habitation and resource development

Challenges of hot, arid and semi-arid environments

Challenges of hot, arid and semi-arid environments include the extremes of temperature, accessibility, water shortages and a lack of resources. For example, this makes farming difficult, and even where irrigation is practised there are high rates of evapotranspiration.

People may wear long, loose, light-coloured clothing to allow air to circulate, reflect insolation and protect their skin against the sun during the day. Turbans provide protection from the sun and can be wrapped around the face to protect against sand storms. Nomadic herders move their flocks to be close to water during dry periods. They do most of their work in the morning to avoid the intense midday heat.

Flash floods occur in hot, arid and semi-arid areas when rainfall intensity exceeds infiltration capacity causing water to flow over the surface. Floods may be brought about by summer convectional rainfall, unvegetated surfaces (thus less interception and infiltration) and concentrations of water in wadis, gullies or channels, for example, the flash floods in Cline Creek and Cave Creek in Arizona, USA in 2014.

Challenges in periglacial and glacial areas

Challenges in periglacial and glacial areas include the possibility of glacial surges, avalanches, landslides, road instability and flooding from glacial melt.

Periglacial environments are characterized by:

- low temperatures that reduce plant growth and make the working environment difficult

- low temperatures that make the provision of water and waste disposal difficult and expensive

- long hours of darkness that limit plant growth and have an adverse impact on human happiness

- thawing of permafrost that may cause subsidence of buildings and infrastructure

- the need for special wheels on vehicles to cope with the ice/snow

- the need to keep some machines running the whole time to prevent shutdown.

▲ Figure C.1.2. A periglacial environment in Thingvellir, Iceland

Many mountainous areas have steep relief, making settlement and economic activity difficult, for example, Brooks Range, Alaska. Limited transport routes make mountainous areas inaccessible. The high relief leads to reduced temperatures, so growing seasons are shorter, and soils are likely to be thin and infertile.

People may adapt to living in cold environments in a variety of ways. For example:

- some people may take vitamin D supplements in winter to make up for a lack of sunshine or use SAD (seasonal affective disorder) lamps to deal with the lack of light in winter

- clothing needs to be windproof as strong winds contribute to wind chill in winter months—heavy clothing, head protection and gloves may prevent frostbite

- buildings may be built on stilts to avoid potential issues with permafrost thaw in summer

- cars need snow tyres to travel over snow and extra heaters to stop diesel fuel freezing due to extreme cold in winter.

The changing distribution of extreme environments over time

Past and present distribution of arid and semi-arid environments

Desertification refers to the spread of desert-like conditions into areas that were previously biologically productive. Much modern desertification is anthropogenic (human induced) but some desertification is natural, due to long-term climate change over the last two million years. For example, the Kalahari sands are mainly fossil desert now covered with acacia and mopane trees.

Past and present distribution of glacial environments

There are a number of interrelated factors that may cause ice ages and glacial phases. These include natural (Milankovitch) cycles, such as:

- the "stretch" in the Earth's orbit around the Sun

> **Test yourself**
>
> **C.3 Describe** one challenge that is similar in hot and semi-arid environments and in cold environments. [2]
>
> **C.4 Explain** how the traditional clothing worn in deserts helps people to adapt to the extreme desert environment. [2]

 Content link
The causes and consequences of global climate change are explored further in units 2.1 and 2.2.

- the "tilt" of the Earth
- the "wobble" of the Earth's axis.

In addition, there are changes in:

- the distribution of land and sea
- reflectivity (albedo)
- tectonic activity
- human effects (anthropogenic impacts).

The past and present distribution of glacial and periglacial environments in the northern hemisphere is related to these very long-term changes in climate.

Test yourself

C.5 Describe how natural cycles affect glaciations.   [3]

C.6 Briefly **explain** how albedo changes lead to changes in the amount of ice present.   [2]

C.7 Suggest how human activities influence glacial advances/retreats.   [3]

C.2 PHYSICAL PROCESSES AND LANDSCAPES

- **Diurnal** – during the course of one day.

You should be able to show how physical processes create unique landscapes in extreme environments:

✔ Glacial processes of erosion, transport and deposition, and landscape features in glaciated areas, including cirques/corries, lakes, pyramidal peaks/horns, arêtes, glacial troughs, lateral, medial and terminal moraine and erratics;

✔ Periglacial processes of freeze-thaw, solifluction and frost heave, and periglacial landscape features, including permafrost, thermokarst, patterned ground and pingos;

✔ Physical and chemical weathering in hot, arid environments, and erosion, transportation and deposition by wind and water;

✔ Hot, arid landscape features, including dunes, wadis, rock pedestals, mesas and buttes.

Concept link

PROCESSES: Natural processes shape environments, with landforms being created in arid and glacial environments. Change occurs as a number of these processes take place simultaneously. There are different rates of change depending on the environment and the feature being formed in the landscape.

Glacial processes of erosion, transport and deposition, and landscape features in glaciated areas

Processes in extreme environments: Glacial environments

Glacial erosion and associated processes consist of a number of different actions.

- Abrasion is often called the "sandpaper effect".
- Plucking (or quarrying) is the ripping out of material from the bedrock.

Types of glacial erosion

- A pyramid peak is formed when three or more arêtes converge.
- A ribbon lake is a long lake found in a trough often formed as a result of damming by moraine.
- A corrie is an armchair-shaped hollow formed by freeze-thaw weathering, abrasion, plucking, and the rotational movement of ice.
- A glacial trough is a steep-sided and relatively flat-floored valley, formed by many cirque glaciers and valley glaciers.
- An arête is a sharp ridge formed by freeze-thaw weathering and the retreat of cirques.

>> Assessment tip

An annotated diagram can be used to show the formation and shape of glacial features.

Glacier transport

A glacier's load can be carried:

*   subglacially—under the glacier

*   englacially—within the glacier

*   supraglacially—on top of the glacier.

Glacial deposition

Terminal moraines mark the maximum advance of a glacier. A good example of terminal moraines is the Franz Josef terminal moraine in New Zealand.

Lateral moraine refers to the moraine carried at the side of a glacier. Much of the material comes from loose weathered rock from the valley sides.

Medial moraines occur when two glaciers meet. The two lateral moraines that meet subsequently flow as one medial moraine in the middle of the enlarged glacier.

Erratics are rocks that have been deposited in an area of different geology. They have been eroded by ice, transported and deposited.

Periglacial processes of freeze-thaw, solifluction and frost heave, and periglacial landscape features

The main periglacial processes include freeze-thaw weathering, solifluction and frost heave.

Freeze-thaw weathering is common and effective at breaking down rocks in many periglacial areas due to the number of freeze-thaw cycles every year. This leads to repeated freezing and thawing, especially in spring and autumn, and leads to the breakdown of many jointed rocks.

Solifluction literally means flowing soil and this is one of the main ways that weathered material is transported downslope. It is a slow movement, occurring at speeds of between 1 cm per day and 1 cm per year. Water can soak into the ground, but only as far as the impermeable permafrost, and it carries particles of soil with it as it flows along the permafrost.

Frost heave is a process in which water that freezes lifts soil particles (peds) and stones within the ground upwards towards the surface.

> **Test yourself**
>
> **C.8 Distinguish** between lateral and medial moraine. [2]
>
> **C.9 Explain** how erratics can be used to analyse glacial movement. [2]

▲ Figure C.2.1. Frost heave of soil

> **Content link**
>
> Freeze-thaw weathering and solifluction are discussed in the context of geophysical hazards in option D.1.

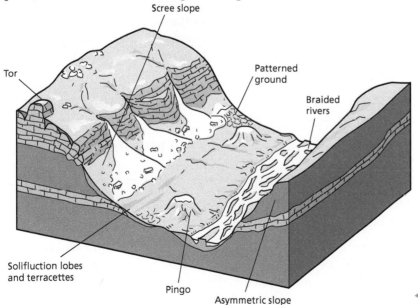

◀ Figure C.2.2. Periglacial landscape features

Permafrost is defined as permanently frozen subsoil (for at least two years). There are three main types of permafrost:

- Continuous permafrost—where the upper limit of the ground effectively remains at the surface throughout the year

- Discontinuous permafrost—where there is significant melting in summer

- Sporadic permafrost—where the permafrost is fragmented.

Pingos are rounded ice-cored hills that can reach heights of over 50 m and widths of over 500 m. There are two main types. Open-system pingos form in valley bottoms where water collects, freezes and forms large ice masses that heave up overlying sediments into domes. Closed-system pingos develop beneath small lakes and lake beds. As permafrost grows, groundwater beneath the lake is trapped by freezing from above and freezing inwards from the edge of the lake basin. The trapped groundwater forms a talik (unfrozen sediments), but is increasingly compressed by the expanding ice that surrounds it. Eventually the overlying sediments are forced upwards, and the talik itself freezes to produce an ice-cored dome.

When there is extensive thawing of ground ice, for example, by fire or due to climate change, an irregular surface of hummocks, shallow depressions and pits may be formed. These are called **thermokarst**.

Patterned ground refers to the regular stone circles and polygons on flatter ground, and stone stripes on steeper slopes, found in areas where there is limited precipitation and limited ice. They can be several metres wide.

Physical and chemical weathering in hot, arid environments, and erosion, transportation and deposition by wind and water

Weathering

Disintegration or insolation weathering is found in hot desert areas where there is a large diurnal temperature range. The rocks heat up by day and contract by night. As rock is a poor conductor of heat, stresses occur only in the outer layers. This causes peeling or exfoliation. In some instances, rocks may split in two.

Erosion

Salt crystallization causes the decomposition of rock by solutions of salt. When water evaporates, salt crystals are left behind. As the temperature rises, the salts expand and exert pressure on rock.

There are two types of wind erosion:

- Deflation is the progressive removal of small material, leaving behind larger materials.

- Abrasion is the erosion carried out by wind-borne particles.

Transportation and deposition by wind and water

Sand-sized particles (0.15–0.25 mm) are moved by three processes: suspension, saltation and surface creep.

Rainfall in dry areas may be irregular and episodic (sporadic) but some desert areas experience occasional heavy downpours that cause flash floods. During a flash flood, the river picks up a large volume of material, which is deposited as the flood subsides.

Test yourself

C.10 **Define** the terms
(a) permafrost and
(b) thermokarst.   [1+1]

C.11 **Explain (a)** the formation of patterned ground and **(b)** the process of frost-heave.   [2+2]

Hot, arid landscape features

Landforms

- Sand dunes are formed where wind speed is high and constant, and where there is a large supply of sand.

- Wadis are dry gullies that have been eroded by flash floods.

- Mesas are relatively large areas of plateau that have become isolated from the main plateau either by erosion or through slope retreat.

- Buttes are isolated peaks, often the remnants of former mesas or plateaus.

- Rock pedestals are rocks that have been eroded closer to the ground (by abrasion) and so have a wider upper part, sometimes resembling a mushroom.

> **Test yourself**
>
> **C.12 Suggest** how the climate of hot, arid and semi-arid environments influences the processes of weathering and erosion. [3]

C.3 MANAGING EXTREME ENVIRONMENTS

You should be able to show the varying power of different stakeholders to extract economic value from extreme environments:

✔ Agricultural opportunities and challenges in arid areas, including the distinction between aridity and infertility, irrigation access, salinization risk and land ownership;

✔ Human and physical opportunities and challenges for mineral extraction in cold environments, including inaccessibility, permafrost and resource nationalism;

   ✔ Case study of one cold environment to illustrate the issues;

✔ Human and physical opportunities and challenges for mineral extraction in arid environments, including inaccessibility and climatic and political factors;

   ✔ Case study of one arid environment to illustrate the issues;

✔ Opportunities and challenges for tourism in extreme environments;

   ✔ Detailed examples illustrating the involvement of local and global stakeholders.

> • **Infertile soils** – soils that are lacking in nutrients or bases.
>
> • **Irrigation** – the addition of water to farmed areas in order to help plants cope with seasonal or permanent shortages of water.

> **>> Assessment tip**
>
> It is dangerous to generalize. The opportunities and challenges in arid areas and cold areas vary with the level of development of the country. HICs (such as the USA and Canada) are more able to deal with such areas than LICs (such as Nepal and Burkina Faso). To avoid this potential pitfall, use case studies to back up your points.

Agricultural opportunities and challenges in arid areas

Aridity refers to a lack of moisture (precipitation less than 250 mm per year), or where there are very high evapotranspiration rates. In semi-arid environments, annual rainfall varies between 250 mm and 500 mm, so there is some possibility for farming, especially where water conservation methods are used.

Desert soils are arid (dry) and sometimes infertile. Infertility refers to the lack of nutrients or bases in the soil. There is a lack of biomass, and so limited organic matter entering the soil, and low weathering rates also reduce the inputs of nutrients.

Salinization may occur in areas where annual precipitation is less than 250 mm. As water evaporates, salts are left behind in the soil making it toxic to many plants.

> **🔗 Content link**
>
> The lack of moisture in arid and semi-arid environments relates to discussions of water security in option A.3 and unit 3.2.

Not all societies have access to irrigation. Even where irrigation is possible, it may lead to problems such as salinization, depletion of groundwater and pollution. Moreover, the extraction of water for irrigation in one area may lead to problems elsewhere.

In some societies in countries such as Mali, Niger and Burkina Faso, land ownership may be limited, especially for women. Formal law, legal systems and social norms often deny women and indigenous people the right to acquire land or to maintain their land.

Nonetheless, there are agricultural opportunities in arid environments. They include:

- nomadism (the traditional way of dealing with insufficient amounts of rainfall and pasture)

- settled farming using groundwater

- irrigation in areas close to rivers or oases

- increased use of drought-tolerant species.

Human and physical opportunities and challenges for mineral extraction in cold environments

Challenges for mineral extraction

Resource development in periglacial environments is hindered by low temperatures that:

- make the working environment difficult

- make it difficult/expensive to provide services such as water and sewage disposal

- mean that some machines have to be kept running the whole time in winter or they shut down.

There are also a number of hazards such as avalanches, rock falls and frost heave. Waste disposal is difficult because of the low temperatures.

In addition, remoteness, inaccessibility, water supply and transport difficulties combine to make mineral extraction in cold environments difficult.

Resource nationalism may also limit access to resources. Denmark remained in control of Greenland's resources until the 1990s. In 2009, Greenland achieved full home rule, including control of natural resources. This resource nationalism was expected to bring great benefits to Greenland. However, the oil price collapse since 2013 has made Arctic oil exploration in Greenland too expensive and not worthwhile.

Opportunities for mineral extraction

The exploitation of periglacial areas for their mineral and fossil fuel resources creates both opportunities and challenges. Resource development can improve the economies of these regions and generate wealth for individuals, communities, TNCs and national governments, but it can also put their fragile environment under pressure and create conflicts among local communities.

Case study: Mining in northern Europe

Large areas of Europe's remaining wilderness areas risk being damaged and polluted as mining companies prepare to develop

Concept link

POWER: Mining resources and agriculture are two examples of land use in extreme environments. The development of these industries may be in conflict with the aims and desires of those already in situ in these areas, and a power dynamic can develop between different vested parties, such as TNCs and local groups. With resource availability declining, governments may seek to enhance their sovereign power by trying to develop their domestic power sources. Government priorities, plus the aims of TNCs, can be pivotal in the decision-making process when places are contested.

northern Finland, Sweden and Norway for uranium, iron ore, nickel, coal, gold, diamonds, zinc and phosphorus. However, these developments could bring permanent damage to the network of rivers, lakes and mountains, as well as to indigenous herders and the tourist industry.

More than one-eighth of Finland has been designated for mining and the Norwegian fertilizer company Yara International plans a 40–60 km² open-cast phosphorus mine near Sokli. Heavy metals such as antimony, copper, cobalt, nickel and chromium from existing mines are contaminating the feeding grounds of reindeer.

Human and physical opportunities and challenges for mineral extraction in arid environments

Opportunities for mineral extraction

Areas of mineral extraction include oil in the Middle East, uranium in Australia and Niger, and copper in Arizona, USA and in the Atacama Desert, Chile.

The opportunities are economic gains from the resources being mined, which provides revenue for the country and for improvement of local infrastructure, investment into local areas, and employment and higher wages provided for local people and/or the migrant workforce.

Challenges for mineral extraction

Challenges include environmental effects on local water supplies, aesthetic changes in the natural landscape, pollution and accelerated erosion. There are also economic and social challenges. In addition, remoteness, inaccessibility and the challenge of attracting a workforce are major problems for some hot, arid environments.

Case study: Mining in Chile

Chile is one of the most important mining countries in the world. Copper accounts for over 60% of Chile's export revenue.

Mining generates tremendous profit for private investors, considerable economic growth potential for Chile, huge potential tax revenues for the Chilean state, and high salaries for many of the workers. The limitations on developing mining include scarcity of water, affordable energy, the rights of indigenous people, and the willingness of the public to tolerate the impacts of mining.

▼ Figure C.3.1. The Atacama Desert is the main focus of Chilean mining.

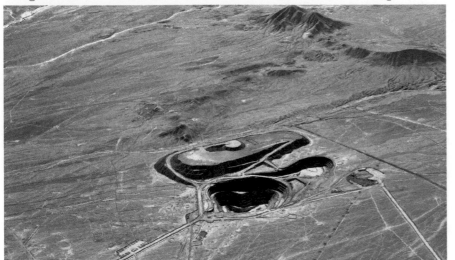

> ### Test yourself
>
> **C.13** Briefly **explain** how permafrost may hinder mineral extraction in periglacial environments. [2]
>
> **C.14 Analyse** problems caused by mineral extraction in cold environments. [10]

▼ Table C.3.1. Top five copper producers ('000 tonnes), 2015

Rank	Country	Tonnes ('000)
1	Chile	5,434
2	China	1,642
3	Peru	1,299
4	USA	1,196
5	Australia	914

Source of data: *The Economist* (2017)

> ### Test yourself
>
> **C.15 Identify** the minerals that are found in Chile. [1]
>
> **C.16 Suggest** why copper is so important for Chile's economy. [3]

Mining has had many negative impacts. The use of groundwater has caused some wetlands to dry out. Mining operations generate considerable dust that can be transported large distances by wind.

Increased competition for water between mine operators, mine workers and indigenous populations is threatening biodiversity. For example, the Cerro Colorado copper mining project pumped 125 litres of water per second out of a protected wetland between 1994 and 2002.

Opportunities and challenges for tourism in extreme environments

In desert regions, opportunities for tourism may be linked to scenery (dunes, salt flats, canyons), wildlife, indigenous culture and outdoor pursuits. Tourism may create many jobs, providing incomes for its workers, major benefits for TNCs and taxes for governments.

Challenges may include extremes of temperature, accessibility, water shortages and a lack of resources to sustain tourism.

The impacts on the natural environment include mass movement, erosion, land degradation, hazards, aesthetic changes, water shortages (and salinization), waste, introduction of exotic species and habitat removal. These can be positive/negative, short-term/long-term, intentional or unintentional.

Case study: Tourism in the United Arab Emirates

- Dubai is the fourth most globally visited city.

- Over 15 million tourists visited Dubai in 2017; 20 million are expected by 2020.

- In 2016, tourism contributed over £5 billion to UAE's GDP—this is expected to rise to over £15 billion by 2026.

- Over 12% of Dubai's population will be working in the travel and tourism sector by 2026.

Tourism in cold environments

Tourism is a major economic activity in cold environments, in particular alpine areas. However, even wilderness areas, such as the Arctic and Antarctica, have seen an increase in visitor numbers, for many reasons:

- The various attractions of its wilderness areas—scenery, wildlife, northern lights

- Improved accessibility

- Improved infrastructure

- Increased personal affluence and leisure time

- The desire for "new" experiences.

In cold or high-altitude regions, opportunities may focus on skiing and other outdoor activities, scenery and experiencing life in indigenous societies. Many governments recognize the potential for tourism and the multiplier effect that it can have. On the other hand, increased tourism can lead to the destruction of the natural environment, pollution, conflict with indigenous peoples and economic leakage. The challenges may also include remoteness, availability of power, access, as well as the mitigation of natural hazards such as landslides and avalanches.

Content link

Link this content with discussions of tourism in option E.

Test yourself

C.17 Outline reasons for the growth of tourism in the UAE.   [3]

C.18 Suggest why the increasing dependence of Dubai on tourism and recreation may lead to problems in the future.   [3]

Test yourself

C.19 Briefly **explain** the growth of tourism in cold environments. [3]

C.20 Analyse two environmental impacts that tourism in cold environments can have.   [6]

C.4 EXTREME ENVIRONMENTS' FUTURES

You should be able to show examples of future possibilities for managing extreme environments and their communities:

✔ The causes, acceleration, consequences and management of desertification including land use, conflict and climate change;

   ✔ One case study illustrating the human and physical dimensions of desertification;

✔ Increasing competition for access to resources in extreme environments, including the role of indigenous groups, civil society organizations, transnational corporations (TNCs) and militia groups;

   ✔ One case study to highlight the issues;

✔ New technology and sustainable development in extreme environments, including greater use of solar power and desalination;

✔ The impacts and management of global climate change in extreme environments, including adaptation by local populations.

> **Desalination** – the removal of salt from seawater to produce freshwater.

The causes, acceleration, consequences and management of desertification

Desertification is defined as the spread of deserts or desert-like conditions into previously productive areas. Desertification is a widespread process, affecting low- and high-income countries alike.

Desertification is the result of many factors:

- Increased pressure on land resources due to migration
- Changes in agriculture
- Increased use of water through more intensive agriculture.

Desertification has resulted in the following:

- Reduced crop yields in irrigated and rain-fed agriculture
- Encroachment of sand-bodies on productive land and settlements.

Wind erosion can be reduced by the construction of wind breaks and by adding organic matter to the soil. Runoff can be captured by the building of small earthen dams (diguettes). Livestock can be kept off an area by putting up fencing. Trees can be planted to provide various functions. They help to bind the soil; they return nutrients to the soil; they reduce flooding; they create shade for animals and people; and they act as wind breaks.

Case study: Desertification in Europe

A fifth of Spanish land is so degraded that it is turning into desert. In many places tourism is making things far worse. Water is pumped from below ground, pulling salt water from the sea into the aquifers.

In some parts of Mediterranean Europe, the degradation is so severe that it has reduced the soil's capacity to support human communities and ecosystems and resulted in desertification.

> **Concept link**
>
> **POSSIBILITIES:** Extreme environments are under threat from socio-economic and environmental pressures. Technological development, as well as adaptation, can ensure that these areas are able to evolve with changing situations, and also enable positive change so that sustainability in these environments is achieved.

Test yourself

C.21 Examine the impacts of desertification. [2+2+2]

C.22 Suggest ways in which desertification can be managed. [2+2]

Increasing competition for access to resources in extreme environments

Case study: The Carmichael coal mine, Galilee Basin, Queensland, Australia

The Galilee Basin in Queensland, Australia contains around 27 billion tonnes of coal. A huge proposed complex of coal mines is planned here, including the world's largest thermal coal project, railway lines and a massive expansion of the Abbot Point port on the Great Barrier Reef. Adani, an Indian firm, is behind the project.

One key environmental impact will be on water, which is limited in Queensland. The Carmichael mine alone will require up to 12.5 billion litres a year. The water table is expected to drop by 50 metres in some places.

Critics argue that mining is unsustainable, while those in favour argue that mining (and agriculture) have always been Australia's primary industries.

New technology and sustainable development in extreme environments

Desalination

Desalination (or desalinization) removes salt from seawater. Seawater is desalinated to produce fresh water fit for human consumption (potable water) and for irrigation.

Due to high energy input, the costs of desalinating seawater are generally greater than the alternatives, but alternatives are not always available and depletion of reserves is a problem in certain locations. In 2015, there were around 18,500 desalination plants operating worldwide, producing 86.8 million cubic metres per day, providing water for 300 million people.

The main criticism regarding desalination and the use of reverse-osmosis technology is that it costs too much.

Solar power

From a relatively small base, the installed capacity of solar electricity is growing rapidly. In 2017, global solar power capacity passed 400,000 megawatts. This amounts to about 1.8 per cent of all global electricity generation. Experts say that solar power has huge potential for technological improvement which could make it a major source of global electricity in years to come.

Advantages	Disadvantages
A completely renewable resource	Initial high cost of solar plants
No noise or direct pollution during electricity generation	Solar power cannot be harnessed during intense storms or at night
Very limited maintenance required to keep solar plants running	Of limited use in countries with low annual hours of sunshine

▲ Table C.4.1. The advantages and disadvantages of solar power

---

Test yourself

C.23 Identify the various stakeholders' viewpoints regarding development of coal mining in the Galilee Basin. [1+1]

C.24 Briefly **explain** (a) the advantages and (b) the disadvantages of developing the mining in the Galilee Basin. [2+2]

 Content link

Desalination in the context of water security is discussed in unit 3.2.

Content link

Solar power, as a renewable energy source, is relevant to the discussions of energy consumption in unit 3.1.

Test yourself

C.25 Briefly **describe** two advantages and two disadvantages of solar energy.   [4]

The impacts and management of global climate change in extreme environments

The Arctic Resilience Report 2016 reported that the effects of Arctic warming could be felt as far away as the Indian Ocean.

Climate tipping points occur when a natural system, such as the polar ice cap, undergoes sudden or overwhelming change that is irreversible. In the Arctic, the tipping points include:

- growth in vegetation on tundra, which replaces reflective snow and ice with darker vegetation, thus absorbing more heat

- increased emissions of methane, a potent greenhouse gas, from the tundra as it warms.

Scientists have speculated for some years that feedback mechanisms could suddenly take hold and change the dynamics of Arctic ice melting from a relatively slow phenomenon to a fast-moving one with unpredictable and potentially irreversible consequences for global warming. For instance, when sea ice shrinks, it leaves areas of dark ocean that absorb more heat than the reflective ice, which in turn causes further shrinkage and so on in a spiral.

Coping strategies (adaptations) in cold environments include:

- changing the type of farming/fishing

- relocating homes away from coastal locations

- increased focus on tourism and recreation.

Content link

The impacts of global climate change are discussed more generally in unit 2.2.

Test yourself

C.26 Describe one example of positive feedback and one example of negative feedback due to potential climate change in the Arctic. [1+1]

C.27 Briefly **explain**, and illustrate, the term "tipping point". [2+1]

QUESTION PRACTICE

Examine the photo on the right, which shows a landform within the Alps mountain range.

a) i) Describe two main features of the landform shown. [2]

ii) Explain the processes of plucking and abrasion. [2 + 2]

b) Briefly **explain** two factors that affect the location of the world's hot deserts. [2 + 2]

Essays

Either: Examine the opportunities and challenges for mineral extraction in cold environments. [10]

Or: Examine the causes and consequences of desertification. [10]

How do I approach these questions?

a) i) This asks you to distinguish, that is, make clear the difference between a glacial trough and a corrie lake.

ii) This asks you to explain. To get two marks you will need to identify the process and then develop the answer with detail/exemplification.

b) You need to identify two factors, and then explain how they affect the location of deserts. Again, to score two marks for each point, you need to develop the answer.

First essay choice:

Good answers are likely to cover both opportunities and challenges in cold environments. If only one is covered, the maximum mark that can be achieved is 6/10. Opportunities and challenges are likely to include a range of environmental, social, economic and political factors. You should include supporting examples/case studies in your answer, and you must include some evaluation to achieve full marks.

Second essay choice:

Good answers are likely to cover both the causes and consequences of desertification. If only one is covered, the maximum mark that can be achieved is 6/10. Causes and consequences are likely to include a range of environmental, social, economic and political factors. You should include supporting examples/case studies in your answer, and you must include some evaluation to achieve full marks.

SAMPLE STUDENT ANSWER

▲ Landform identified

▲ Two features described

a) i) The landform is a glacial trough (U-shaped valley).

It has very steep valley sides, and a concave valley floor.

Marks 2/2

▲ Valid point

▲ Development point

▲ Valid point

▲ Valid development

ii) Abrasion is the scratching and scraping of rocks like sandpaper and it is most effective when the load carried by the glacier is stronger/tougher/more angular/sharper than the rock over which it moves In contrast, plucking is the ripping of rock from underneath the glacier. It occurs on jointed rocks where meltwater can get into joints, freeze and pluck or rip the rock away from the solid rock below.

Marks 4/4

▲ Valid point

▲ Valid explanation

▲ Valid point

▲ Valid development

b) Some hot deserts such as the Namib Desert are found by a coast with cold currents. The cold current prevents much rain from forming, although there may be fog in coastal areas. Other hot deserts are found at great distance from the sea, e.g. Central Sahara Desert. This means that any water-bearing winds that there could have been have lost their moisture by the time they get to the more continental interiors.

Marks 4/4

Essays

Either: Examine the opportunities and challenges for mineral extraction in cold environments.

▲ Good point, lots of information

▲ Good range of resources and areas

▲ Potential and impacts of resource development

▲ Introduces a range of advantages with some support

There are many opportunities for mineral extraction in cold environments. The Arctic is believed to contain up to 90 billion barrels of undiscovered oil (13% of undiscovered oil), 30% of undiscovered natural gas and 20% of undiscovered natural gas liquids. In addition, there are significant deposits of iron ore at Kiruna (Sweden) and gold in Alaska. So, there are many potential benefits in terms of job creation, increased wealth for mining companies and tax revenues for countries.

However, there are many disadvantages for companies trying to extract minerals in cold environments. The low temperatures make working conditions difficult. Some oil workers in Siberia are paid seven times as much as oil workers in other parts of Russia. The low temperatures make it difficult to provide services such as water and sewerage disposal. During winter there are very long nights – in some places more than twenty hours of night time. Vitamin D deficiency is a real problem for many people.

▲ Disadvantages

▲ Quantification

▲ Explanation

▲ Range of valid points with some explanation

The permafrost makes it very difficult to operate due to thawing (by heat) which can lead to subsidence. The weight of vehicles may cause pressure on the permafrost and cause it to melt. In order to build roads and buildings, foundations need to be built into a thick gravel pad. However, the extraction of gravel removes valuable fish-spawning habitats from rivers.

▼ Fair points but would be good to say "where" exactly.

▲ Simple explanation

▲ Further explanation

▲ Solution to problem

▲ Negative environmental impact

There are also many hazards in cold environments, including avalanches, rock falls, icings and frost heave. These make mining activities more difficult. A further problem is that any pollution that is produced, such as oil leaks, take a very long time to decompose, due to the low temperatures. Most recently, there has been climate change which is causing the temperatures of cold environments to rise – this is not necessarily a good thing as it can increase the amount of permafrost melt.

▲ Good explanation of problems

▲ Geophysical hazards

▼ Negative impact—but why?

▲ Well explained repetition from above

▲ Recent changes

Covers a range of advantages and disadvantages. Disadvantages given greater attention. Some support but not very detailed. No conclusion given.

Marks 8/10

Or: Examine the causes and consequences of desertification.

Approximately 25% of the Earth's land surface experiences desertification. Desertification is caused by complex interactions among physical, biological, political, social, cultural and economic factors. It can result from prolonged drought, sustained high temperatures, forest fires, reduced vegetation cover and a range of human activities such as over-grazing, over-cultivation, reduced periods of fallow.
Desertification can result in a decline in soil fertility, falling land productivity, reduced crop yields, poverty and malnutrition. Soils may experience loss of nutritive matter (due to agricultural over-exploitation); loss of soil surface due to wind and rain erosion, particularly due to the loss of

▲ Complex issue

▲ Range of causes

▲ Identifies a range of natural and human causes of desertification

▲ Impacts

▲ Good points made about soils

▲ Impacts on water

▲ More environmental impacts

▲ Now has impacts on people

▲ Impact on food security

▲ All valid points

vegetation; soil pollution (due to the excessive use of chemical fertilizer); the effects of compression and the encrusting of the soil surface (due to heavy use of agricultural machinery). Water resources become scarce and groundwater tables may fall. Ironically, floods may occur when it rains due to the limited vegetation cover. Plant and animal species may become extinct – there is a reduction in biodiversity with knock-on effects for the availability of food from dry areas.

Desertification affects nearly one billion people. It can lead to a vicious circle of poverty and increased exploitation of a dwindling resource. Falling food production not only leads to an increase in hunger and malnutrition, it leads to falling incomes for farmers and reduced revenues for national governments. It may make them turn to food aid, creating a dependency upon imported foods. It may lead to increased migration, and increasing pressures in urban slums.

Generally a very good account—a case study or named examples would be beneficial.
Marks 8/10

D GEOPHYSICAL HAZARDS

The human and natural worlds face a number of risks, and geophysical hazards are a constant threat in many parts of the world. Dynamic tectonic processes ensure that places at varying levels of development have to cope with the impact of volcanoes and/or earthquakes as well as different types of mass movement.

You may have already studied other risks such as extreme climatic events as part of Unit 2, so you should be familiar with the conceptual connections such as the processes involved, the effect on different places, the power of different stakeholders in coping with the risk and the events, and the possibilities to create resilience.

You should be able to show:

✔ how geological **processes** give rise to geophysical events of differing type and magnitude;

✔ how geophysical systems generate hazard risks for different **places;**

✔ how the varying **power** of geophysical hazards can affect people in different local contexts;

✔ how future **possibilities** can lessen human vulnerability to geophysical hazards.

D.1 GEOPHYSICAL SYSTEMS

You should be able to show how geological processes give rise to geophysical events of differing type and magnitude:

✔ Mechanisms of plate movement including internal heating, convection currents, plumes, subduction and rifting at plate margins;

✔ Characteristics of volcanoes (shield, composite and cinder) formed by varying types of volcanic eruption; associated secondary hazards (pyroclastic flows, lahars, landslides);

✔ Characteristics of earthquakes (depth of focus, epicentre and wave types) caused by varying types of plate margin movement and human triggers (dam building, resource extraction); associated secondary hazards (tsunami, landslides, liquefaction, transverse faults);

✔ Classification of mass movement types according to cause (physical and human), liquidity, speed of onset, duration, extent and frequency.

- **Convection currents** – the transfer of heat via movement of magma in the Earth's crust.

- **Subduction** – when a tectonic plate is forced underneath another tectonic plate into the mantle at a convergent plate boundary.

- **Rifting** – the creation of a crack or fault line in the Earth's crust as the lithosphere is extended and stretched.

- **Pyroclastic flow** – a rapidly moving mixture of hot gases, rocks and lava that is produced when a volcano erupts.

- **Lahar** – a flow of volcanic debris that has mixed with water and mud.

- **Liquefaction** – when a saturated land surface changes composition, moving from a solid to a liquid temporarily due to seismic activity in the Earth's crust.

- **Transverse fault** – when rocks move in opposite directions to one another creating tension and a release of seismic energy.

Mechanisms of plate movement

The Earth is a system that is constructed from a series of layers. Each of the layers has a different composition, and it is the interaction between these layers that drives the processes of tectonic movement within the Earth's crust. In particular, convection currents within the mantle affect the overlying lithosphere and this ensures that tectonic plates converge, diverge or compress against each other.

≫ Assessment tip

You should be able to describe and explain the processes that take place at convergent, divergent and transform plate boundaries and ensure that you include appropriate terminology. For example, tectonic plates move apart at divergent plate margins, such as the North American plate and the Eurasian plates, due to convection currents in the mantle. This allows magma to rise and solidify, which means that the sea floor spreads as the plates move apart under the Atlantic Ocean. Underwater shield volcanoes can be formed, which may reach above the water level over time due to further eruptions.

Test yourself

D.1(a) State two types of volcano.   [2]

(b) Describe the differences in terms of the lava emitted from the two types of volcano that you named in part (a).   [2]

When the different types of movement occur, they result in the formation of various landforms and the creation of hazard events. For example, when oceanic crust converges against continental crust, the oceanic tectonic plate (a denser rock type) is subducted or forced underneath the continental plate into the mantle. This creates friction between the two plates and eventually leads to seismic energy being released. Subsequently an earthquake occurs, the land mass on the continental plate is forced to compress and fold mountains are formed. In addition, plumes of magma are formed as the oceanic plate melts in the mantle. These plumes, under pressure, will make their way to the surface via weaknesses in the continental plate. Upon reaching the surface, a volcanic eruption will occur.

Figure D.1.1 shows these different types of plate movement and how they connect within the Earth's geophysical system:

▼ **Figure D.1.1.** Different types of tectonic plate movement

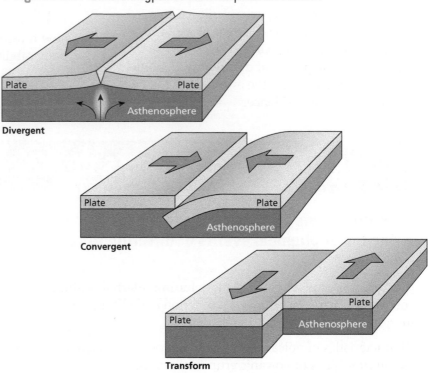

Note the presence of the continental rift zone in the diagram of divergent plates, which shows how a continental plate is being stretched.

▼ **Figure D.1.2.** Mass movements in the Dolomites, Italy

Characteristics of volcanoes

Composite volcanoes, shield and cinder volcanoes have different characteristics in terms of their shape, the type of eruption and the hazards created during and after an eruption. The hazards can be classified into primary (e.g. lava flows, pyroclastic flows, volcanic gases, ash fall) and secondary (e.g. lahars, landslides, flooding, tsunamis, food shortages).

▼ Figure D.1.3.

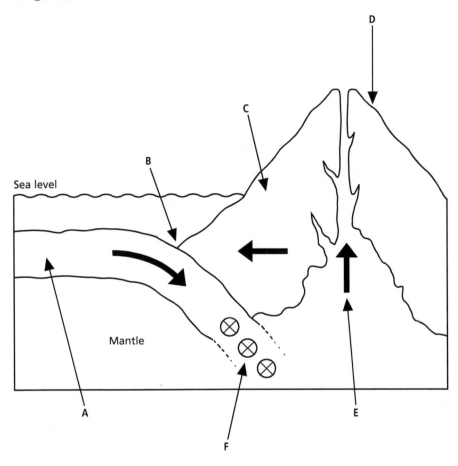

> **Test yourself**
>
> **D.2** Copy figure D.1.3 and **annotate** to describe or explain the process that is taking place at each letter. [6]

> **Assessment tip**
>
> As with other examination questions for the optional themes on paper 1, this type of question provides an opportunity to draw an annotated diagram. Your diagram and your annotations need to be clear, and they should explain the processes and features.

> **Test yourself**
>
> **D.3 Explain** the formation of landforms at divergent plate margins. [2+2]

Characteristics of earthquakes

The point on the Earth's surface from where seismic energy emanates is called the epicentre. The true centre of the earthquakes is in the crust, called the focus. There are different types of seismic waves. P-waves compress and expand the ground like an accordion, affecting both liquid and solid surfaces. S-waves only travel through solid material and not liquid; they move from side to side as well as up and down. Surface waves form when P- and S-waves reach the surface, hence the name. Surface waves fall into two categories: Rayleigh waves which roll along the surface like a wave, and Love waves which move the surface from side to side.

Classification of mass movement types

Human activity can increase the risk of seismic activity. The presence of fracking has been an issue in some areas as it has resulted in minor earthquakes, such as in parts of Oklahoma, USA. Dams are often built in areas where tectonic movement has taken place, and the valley is flooded. Water pressure in cracks behind the dam will increase, which creates weakness and instability in the underlying rock beneath the dam.

Content link
Option A.4 examines the pros and cons of constructing dams.

Mining can also create instability in the Earth's crust. Earthquakes can occur due to the extraction of fluids underground, such as water when mining coal for example, which creates subsidence and movement in the crust.

Figure D.1.4 shows the different types of mass movement.

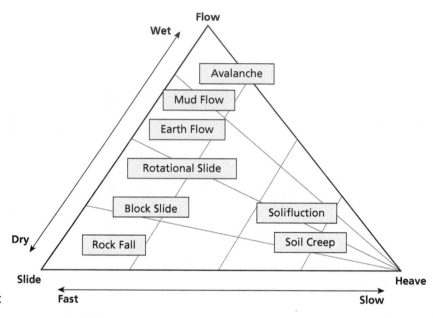

▶ Figure D.1.4. Types of mass movement

Weathering also creates instability and enables different types of mass movement, and seismic energy causes unpredictability in a slope. Weathering can be categorized as mechanical, chemical or biological.

▼ Table D.1.1. Categories of weathering

Mechanical	Chemical	Biological
Freeze–thaw	Hydrolysis	Chelation
Pressure release	Oxidation	
Salt crystallization	Hydration	
Thermal expansion/exfoliation	Solution	
	Carbonation	

Test yourself

D.4 Compare and contrast flows and slides.   [2+2]

 Content link
Connect this information with the effects of increasing temperature due to global warming described in unit 2.1.

Freeze–thaw weathering can create instability due to the expansion and contraction when water freezes and melts in cracks, and can lead to rock falls on dry, steep slopes. Where there is more moisture to lubricate the surface, a rock slide can occur.

A rock avalanche is the most rapid type of mass movement and can often travel down hollows where previous avalanches have taken place.

A rotational slip is concave in shape and normally occurs when a weak rock type such as clay becomes saturated. The clay will then slip as it gets heavy in conjunction with gravity.

Flows can be differentiated according to the size of the particles involved in the movement. In a debris flow, more than 50% of the particles are coarser than sand, whereas in a mud flow more than 50% of the particles are finer than sand. A lahar is an example of a mudflow.

Creep is a slow movement that is caused due to the expansion and contraction as well as heating and cooling of soil, as moisture is absorbed and then released via evaporation.

A slump occurs when rock or soil falls in blocks independent of a curved slip plane underneath.

Solifluction is the downwards movement of soil over a permanently frozen subsurface, and it is common where upper permafrost melts and moves over more frozen permafrost underneath.

 Content link

Mass movements occurring in extreme environments is explored in option C.2.

Test yourself

D.5 Explain why some types of mass movement cause more problems for people than other types of mass movement. [4]

D.2 GEOPHYSICAL HAZARD RISKS

You should be able to show how geophysical systems generate hazard risks for different places:

✔ The distribution of geophysical hazards (earthquakes, volcanoes, mass movements);

✔ The relevance of hazard magnitude and frequency/recurrence for risk management;

✔ Geophysical hazard risk as a product of economic factors (levels of development and technology), social factors (education, gender), demographic factors (population density and structure) and political factors (governance);

✔ Geographic factors affecting geophysical hazard event impacts, including rural/urban location, time of day and degree of isolation.

Magnitude – the size and extent of a geophysical hazard.

Risk – the probability of a hazard event causing harmful consequences such as threats to life, property and infrastructure.

Governance – the ability for the local or national authorities to mitigate the risk from a geophysical hazard.

The distribution of geophysical hazards

Earthquakes and volcanoes tend to be located along the edges of major tectonic plates and as such they have a linear pattern. Mass movement linked to an earthquake or volcanic eruption will also take place in these same locations as well as in areas that are mountainous and areas where there have been land-use changes resulting in instability.

The relevance of hazard magnitude and frequency/recurrence for risk management

The magnitude of earthquakes and volcanoes can be measured, and events with a larger magnitude tend to be more infrequent. The Richter scale, created by Charles Richter in 1935, is designed to assess the magnitude of earthquakes. It is a logarithmic scale and measures factors of 10. For example, a quake that measures 4.0 is ten times more powerful than a 3.0.

Test yourself

D.6 Describe the distribution of different types of volcanoes. [3]

Concept link

PLACES: Risks for people and places are increased due to a combination of natural and human factors which vary from place to place. The natural reasons tend to relate to relative distance to a volcano, a fault line, or a slope, whereas human reasons incorporate a number of socio-economic and political factors, quite often relating to the human development status of a place.

Test yourself

D.7 State the relationship between the magnitude of a hazard event and the frequency of its occurrence.   [2]

D.8 Describe a scale that is used to measure the magnitude of the extent of a volcanic eruption.   [3]

The Mercalli scale is used to measure the damage caused by an earthquake. More recently, in 1979, the Moment Magnitude scale was developed. It uses a greater range of variables than the Richter scale to assess the amount of seismic energy, specifically the movement of rock along a fault line and the area where a surface is ruptured. A seismometer is used to detect seismic waves.

The Volcano Explosivity Index (VEI) is used to measure volcanic eruptions. It is a complex measure that includes the amount of material emitted and the height of an ash column during an eruption. The eruption of Mount St Helens in 1980 was VEI 5 and Mount Pinatubo in 1991 was VEI 6. The Index is logarithmic, similar to the Richter scale, and the top of the Index is VEI 8, which is one million times more explosive than a VEI 2. The United States Geological Survey (USGS) has stated that a VEI 5 normally happens once every 10 years and is much more infrequent than a VEI 3 (which occurs several times a year). A VEI 1 will only emit 0.0001–0.001 km^3 of material during an eruption, while VEI 2 will emit 0.001–0.01 km^3 of material, 10 times more than VEI 1.

Geophysical hazard risk

A range of factors can increase the risk faced by people in relation to geophysical hazards. Some of these factors are:

- **Economic:** The level of economic development will dictate the quality of the emergency services that respond to a hazard event, the ability to fund an early warning system and the stabilization of a slope such that it is less susceptible to mass movement.

- **Social:** Educated people are more likely to understand warnings released by governments, for example, how to prepare for an earthquake. When an earthquake occurs, females are more at risk than males, as gender inequalities are exacerbated during natural disasters. For example, for every one male that died in the Indian Ocean tsunami in 2004, four females died. This was partly due to inequalities in education, which meant women were less likely to know how to swim and climb trees for safety, accordingly to a study by Oxfam. Universities in conjunction with the government may undertake research and monitoring in order to improve the knowledge of hazard risk within a country.

- **Political:** Policy decisions taken by governments often ensure emergency services are funded and prepared for a hazard event, as well as providing appropriate lines of communication in order to warn people. The rules and regulations that govern construction will also be an important factor, especially in urban areas where there is a greater population than in rural areas. For example, all new buildings in Chile have to be able to withstand a 9.0 magnitude earthquake.

Test yourself

D.9 Suggest how disparities in education can increase the risk from geophysical hazards.   [4]

Geographic factors affecting geophysical hazard event impacts

There are also other factors that can increase risk, such as the day of the week and the time of day when an event occurs, as well as the distance between a centre of population and the location of the eruption, earthquake or mass movement.

Test yourself

D.10 "Social and economic factors are the sole causes of the impacts from geophysical hazard events." **To what extent** do you agree with this statement? [10]

Assessment tip

This may be an essay question rather than a short answer response, and the command term "to what extent" should prompt you to include evaluation. Here, for example, there is agreement and disagreement with the statement in the question.

D.3 HAZARD RISK AND VULNERABILITY

You should be able to show how the varying power of geophysical hazards can affect people in different local contexts:

- ✔ Two contemporary contrasting case studies each for volcanic hazards, earthquake hazards and mass movement hazards;

- ✔ For each geophysical hazard type, the case studies should develop knowledge and understanding of:

  - ✔ geophysical hazard event profiles, including any secondary hazards;

  - ✔ varied impacts of these hazards on different aspects of human wellbeing;

  - ✔ why levels of vulnerability varied both between and within communities, including spatial variations in hazard perception, personal knowledge and preparedness.

• **Vulnerability** – the susceptibility of a community to the impacts of a hazard event.

Case study: Volcanic eruption at Volcán de Fuego, 2018

Hazard event profile: Volcán de Fuego, a composite volcano in Guatemala, is one of three major volcanoes located close to the former capital, Antigua. It is known for being constantly active and with Vulcanian eruptions. At the beginning of June 2018 it erupted, producing lava, toxic gas, ash clouds, lahars and pyroclastic flows. The eruption was unexpected due to a lack of information and the fact that just one seismometer had been monitoring the volcano. More material was emitted than had been seen since 1974.

Impacts: The eruption immediately affected almost 13,000 people living in the vicinity of the volcano. There was an evacuation, but not before 110 people died. People were given temporary shelter. The UN Refugee Agency stated that more than 1.7 million people would be affected in the weeks following the eruption. La Aurora International Airport in the capital, Guatemala City, was closed for several days due to the amount of ash in the atmosphere. A local organization estimated nearly 3000 people may have died in total.

Vulnerability: Guatemala is a poor country and the area around the volcano is predominantly rural. The volcano erupts approximately 15 times a year, so the people are used to periods of uncertainty.

Concept link

POWER: While some geophysical hazards can be predicted, the power of geophysical hazards can ensure that places still face a tremendous amount of devastation. Earthquakes in particular can severely impact places, especially if it is challenging to implement measures to reduce the vulnerability of people living there.

▼ Figure D.3.1. An eruption occurs at the Volcán de Fuego, Guatemala

Preparedness: The volcano has been monitored and volcanologists from the National Institute of Seismology, Volcanology, Meteorology and Hydrology of Guatemala (INSIVUMEH) informed the government at 6am that there was a risk of an eruption with pyroclastic flows to follow. But the warnings from the government were only issued after the eruption began at 3pm. The rural settlement of El Rodeo was covered with ash and volcanic rock, and over a hundred people died. Residents were critical of the authorities for not warning them about the impending disaster. Locals in these rural communities are given some limited training regarding evacuation procedures. If the government does not issue an order to evacuate, the responsibility lies with these groups to decide whether to leave. However, some people were evacuated, namely people at an upmarket tourist resort called La Reunion after the owners saw the information from INSIVUMEH.

Case study: Earthquake in Nepal, 2015

Hazard event profile: Nepal has experienced many earthquakes in its history, since it lies in an active seismic zone. The Indian plate is being subducted underneath the Eurasian plate, which creates tension, and eventually pressure is released in the form of seismic energy. The epicentre for the 7.8 earthquake in 2015 was quite shallow at a depth of 7 miles. It was 48 miles from Kathmandu and occurred just before midday. A series of aftershocks followed, including a 6.7 shock the day after the original earthquake.

Impacts: Almost 9,000 people were killed, 20,000 were injured and overall 8 million people were affected. Some 600,000 homes were destroyed and over 250,000 homes were damaged. Water and electricity were not available in many places a week after the earthquake. An avalanche occurred on Mount Everest which engulfed the south base camp and killed 19 people.

Vulnerability: Nepal has a high population density living in a mountainous region, so it is challenging to reach areas that are affected by earthquakes or landslides. Buildings are poorly constructed and the people are poor, with almost half of the population struggling to buy food. Kathmandu is built on a former lake bed, so it is susceptible to seismic waves. During the 2015 earthquake, liquefaction occurred.

Preparedness: Fortunately, hospitals in Kathmandu had been retrofitted in order to withstand earthquakes, and staff had received training in a hospital emergency training plan. But the authorities were surprised at the magnitude of the earthquake and rural areas were much less prepared than urban areas.

Case study: Mass movement in Rwankuba, Rwanda 2018

Hazard event profile: On 6 May 2018 a major mudslide occurred in Rwankuba, Rwanda. It happened after a period of heavy rain which fell continually throughout the previous night.

Vulnerability: Rwanda is one of the most densely populated countries in Africa with many of its population living at the foot of one of several mountains in the country. Due to deforestation, the slopes of these mountains are susceptible to mass movement and so people are at risk. These rural communities depend on agriculture, and there is a lack of drainage on the slopes. Prior to 6 May, a number of other landslides had taken place with deaths and injuries recorded. The west of the country had received above average rainfall. Population density is growing due to increased life expectancy and a relatively high total fertility rate (approximately 3.7 in 2018).

Preparedness: The government had asked people to leave the areas that were at risk, but they did not follow that advice and decided to stay in their villages. Rwanda was one of the first countries in the world to adopt the Sendai Framework Monitor, which reduces the risk of disasters by implementing local development plans, particularly in relation to flooding, landslides and lightning strikes.

Impacts: 18 people were killed, 12 were injured in the villages of Rubazo, Bisesero and Gatsata, and 300 people were made homeless.

Test yourself

D.11 Using a specific case study, **explain** the causes and consequences of a rapid mass movement. [3+3]

D.12 Suggest reasons why communities often underestimate the probability of a tectonic hazard event occurring. [4]

D.13 Explain the ways in which vulnerability to a geophysical hazard can be reduced. [2+2]

>> **Assessment tip**

You should ensure that the example you have chosen for question D.11 represents rapid mass movement, and while the human causes should be explained, natural factors should also be included. The consequences could be categorized into the economic, social and environmental impacts. The consequences should be explained.

>> **Assessment tip**

Geophysical hazards include volcanic eruptions, earthquakes and mass movements, and each of these can be discussed. An earthquake will provide more opportunities for developing explanations.

>> **Assessment tip**

Try to discuss examples and case studies that have taken place in your lifetime. There have been a range of hazard events that have occurred, and you should be in a position to discuss eruptions, earthquakes and mass movement in detail.

Better answers should present a discussion of the relative damage caused by the initial hazard event and that caused by secondary effects; concluding remarks may agree or disagree with the statement.

D.4 FUTURE RESILIENCE AND ADAPTATION

• **Resilience** – the capability of a place to recover from the impacts from a geophysical hazard.

• **Slope stabilization** – ensuring a slope is not susceptible to mass movement by implementing a strategy.

You should be able to show how future possibilities can lessen human vulnerability to geophysical hazards:

✔ Global geophysical hazard and disaster trends and future projections, including event frequency and population growth estimates;

✔ Geophysical hazard adaptation through increased government planning (land-use zoning) and personal resilience (increased preparedness, use of insurance and adoption of new technology);

✔ Pre-event management strategies for mass movement (to include slope stabilization), earthquakes and tsunami (to include building design, tsunami defences), volcanoes (to include GPS crater monitoring and lava diversions);

✔ Post-event management strategies (rescue, rehabilitation, reconstruction), to include the enhanced use of communications technologies to map hazards/disasters, locate survivors and promote continuing human development.

 Content link
You will study megacities in unit 1.

Test yourself

D.14 Describe the distribution of areas at very high risk. [3]

D.15 Identify three highly populated areas that are in areas at very high risk. [3]

D.16 The world's fastest-growing cities are located in Niger, Burundi, Nigeria, Burkina Faso, Tanzania, Mali and Angola. Using figure D.4.1, **state** whether this will mean more people are susceptible to seismic hazard risk. [1]

Global geophysical hazard and disaster trends and future projections

The following map shows those areas at risk due to seismic activity and also the location of megacities.

▼ **Figure D.4.1.** Megacities (circled), and areas at risk due to seismic activity

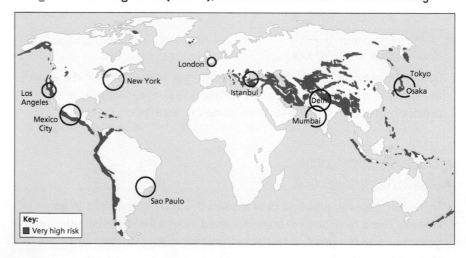

Assessment tip

It should be noted that the information in question D.16 only mentions growth rather than the total number of people living in these cities and also does not state the location of these cities within each country. Some parts of countries are more at risk than others.

Geophysical hazard adaptation

Historical analysis can help to inform city authorities about areas that are more at risk from earthquakes, although the unpredictability of this type of hazard makes land-use zoning a challenge. However, zoning can be used to protect areas from landslides and volcanic eruptions. Areas at risk from mass movement and eruptions can be zoned so that people do not live and work in those areas. The island of Montserrat has a number of areas that people are not allowed to visit unless the level of seismic activity is low.

Pre-event management strategies

There is a range of strategies that can be implemented in order that the future impacts from a geophysical hazard can be reduced. Quite often these strategies are only implemented after a hazard event has taken place, since the risk may not have been known previously or the impacts were unforeseen. In addition, the level of knowledge about the different strategies may have been limited or the expertise not available. One example is modifying construction codes so that buildings are built to withstand a higher magnitude earthquake. This involves modifications such as steel reinforcement, base isolators, movable hydraulic joints, strategies to reduce building shaking, shatterproof glass, deep foundations. By including these things when designing buildings in order that seismic energy is absorbed, they should not collapse or suffer damage during a seismic event.

Test yourself

D.17 Discuss how building design can be the most effective way for people to reduce their vulnerability to earthquakes. [2+2+2]

In terms of secondary hazards, a warning system can alert places about an impending tsunami. Sensors on the sea bed send data to buoys on the ocean surface which is then transmitted via satellite to warning centres.

A slope can be modified in order that the land does not slip, for example by stabilizing it with a metal mesh to prevent rock falls.

Table D.4.1 summarizes the other ways in which a slope can be stabilized.

▼ Table D.4.1. Strategies for stabilizing slopes

Strategy	How it works
Removing groundwater	This can be achieved by using pipes to remove water or by pumping out the water. Different soil types can make it more challenging to achieve this.
Improving surface drainage	The removal of areas that will allow water to accumulate and the drainage of water from a surface via the use of pipes.
Removing material	The excavation of soil and rock at the top of a slope can reduce the pressure that may cause a landslide.
Installing piles	Metal beams are installed in the ground that is underlying the unstable soil and rock in order to create stability.
Constructing walls	A wall built from concrete, rock or logs is often used with the installation of piles if material slips between the piles.
Removing unstable material	If the soil is liable to move, it is replaced with material that is less liable to move. This could be soil or rock that is less prone to weathering.
Afforestation/reforestation/ planting vegetation	The planting of vegetation can help to stabilize a slope and remove moisture from the soil.

Volcanoes are monitored with GPS technology that monitors seismic movements and the release of volcanic gases.

Concept link

POSSIBILITIES: Pre- and post-event strategies can be implemented involving a range of different stakeholders and the use of technology in order to reduce the risk and vulnerability of places. These possibilities not only include financial input, but also the education of citizens who also need to take ownership of responsibility for safeguarding themselves.

≫ Assessment tip

If the command term for this question was "evaluate" then it would be possible to offer an alternative view by stating that land-use zoning, early-warning systems, evaluation planning and increased education would represent more effective ways of reducing vulnerability.

▼ Figure D.4.2.
Earthquake-resistant pipeline

▼ Figure D.4.3.
Slope stabilization in Brunei

Post-event management strategies

Case study: Earthquake in Nepal, 2015

Rescue: The government immediately began to search for people in collapsed buildings. International charities such as Christian Aid, Oxfam and the Red Cross sent teams to help and the Red Cross opened a blood bank in Kathmandu. Facebook users notified friends and family that they were all right via a Safety Check feature which over 8.5 million people used. Over US$15 million was donated using Facebook. Google's Person Finder used crowdsourcing information to log people who had been rescued in order to reconnect them with family members. The Ushadidi internet-based platform was used to collect information about the immediate needs of people affected. Helicopters were used to assess the damage and help rescue people.

Rehabilitation: Temporary shelters were provided for those made homeless. Markets were restored in order that access to food improved as summer crops were harvested. Temporary schools made of bamboo and tarpaulin opened after a month and helped children process the traumatic event.

Reconstruction: Those whose homes were destroyed received 200,000 rupees, and people affected were given access to an 'Earthquake Victim Special Loan' of 25,000 rupees in the Kathmandu Valley. A plan was proposed to ensure that buildings were resistant to earthquakes and a National Reconstruction and Rehabilitation Fund (NRRF) was created to raise US$2 billion to fund the reconstruction. A year after the earthquake, towns and villages outside of Kathmandu remained severely damaged with debris still present. This was due to the amount of legislation and government conditions that had to be satisfied. Two years after the earthquake, only 5% of homes had been rebuilt and many school buildings were still only temporary structures.

Additional post-event responses: Families received 40,000 rupees to cremate a family member, whilst 25,000 rupees was given to those requiring hospital treatment and an amount was provided to those that were made disabled. External funding of approximately US$4.1 billion came via grants and loans, although the government was very slow in releasing this money for projects. Violent clashes took place as people objected to a rapidly introduced first Constitution for the country, and these had to be dealt with by the authorities while politicians focused on the Constitution, which delayed the creation of a National Reconstruction Authority.

🔗 Content link

The use of the Ushahidi platform in building resilience is explored further in unit 6.3.

Test yourself

D.18 Explain how a place is able to increase preparedness before a geophysical hazard event involving mass movement. [2+2]

D.19 To what extent are volcanic eruptions easier to predict but more difficult to respond to than earthquakes? [10]

≫ Assessment tip

A "to what extent" type question, such as D.19, would require an essay or extended response since it requires evaluation. You have the option to discuss the causes, the effects and the responses alongside a range of variables such as building design, early warning systems and other forms of being prepared and responding. Time management is therefore very important and allocating a set time of approximately 22–25 minutes for your essay response at the end of each option is imperative.

Both volcanoes and earthquakes must be discussed, and you should ensure that your paragraphs have a focus. For example, one paragraph could explain how volcanic eruptions are more measurable and predictions can be made based on the changes and release of gases. Your next section could then discuss the responses.

Examples of how to predict earthquakes and volcanic eruptions should be included.

Examine the figure on the right, which shows the location of 5,741 rainfall-triggered landslides from 2007–2013, in blue.

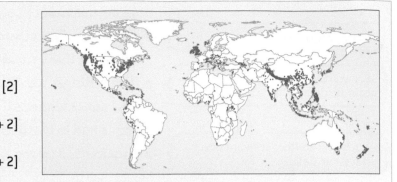

a) **Describe** the global distribution of landslides. [2]

b) **Suggest** two social factors that can increase a person's vulnerability to earthquakes. [2 + 2]

c) Using an example, **explain** two negative impacts of a volcanic eruption. [2 + 2]

Essay

Examine the role of technology in increasing the resilience of places that are susceptible to geophysical hazards. [10]

How do I approach these questions?

a) 2 marks are available for this question and therefore two separate points are required.

b) When social factors are discussed in Geography, they relate to health and education primarily. You should state two factors that are distinct from each other and then you explain each of them, making a clear link to increasing risk. You may want to also include an example at the end of each point.

c) Four marks are awarded for two separate points, with development of these points. Bear in mind that there are short-, medium- and long-term effects. Avoid discussing positive effects, such as fertile soil, as this is irrelevant to the question. Ensure that you refer to an example, as stated in the question. As the command term is "explain", the two impacts stated must be developed to show how people and places were affected negatively.

Essay

The use of technology has helped places to increase their resilience to a range of geophysical hazards and this involves both primary and secondary hazards. You will have studied a range of examples in the final part of this unit for different geophysical hazards. The command term is "examine" which means that you should discuss an assumption that technology is increasing resilience in places by perhaps findings issues. Key concepts such as "possibility" and "power" could be discussed as well as spatial interaction since geophysical events can cover large areas and information can be shared between places.

SAMPLE STUDENT ANSWER

a) Landslides tends to be found along linear patterns such as at the edge of plate boundaries with a large concentration in the Philippines, west coast of the USA, and in the UK.

Marks 2/2

b) The education of people in an area can help people cope when an earthquake strikes. This is because people will have been trained in how to protect themselves and also where to evacuate to when escaping a building. Therefore places where people have not been educated in this area are at a greater risk and are more vulnerable. Government funding into early warning systems will ensure that people have information to be able to evacuate from buildings. Mexico City is an example of a place that receives early warnings from Mexico's Pacific coastline.

The first point is valid and is developed but the second point, while including accurate knowledge, refers to an economic factor rather than a social factor.

Marks 2/4

▲ Appropriate example

▲ 1 mark: relevant effect with accurate detail for the example

▲ 1 mark: another relevant effect with some detail

c) The eruption of Volcán de Fuego caused significant effects to the people of Guatemala. A primary effect was the loss of life to the people living and working close to the volcano and 100 people died. Villages were buried in ash and mud which covered houses, and people also needing rescuing from these houses and 4,000 people had to live in shelters in the days and weeks after the eruption took place.

2 marks for the two negative effects given and 2 further marks for the development of these points.

Marks 4/4

Essay

Examine the role of technology in increasing the resilience of places that are susceptible to geophysical hazards.

▲ Knowledge demonstrated

▼ Mass movement should also be included

▼ This term should be defined in the introduction to show knowledge

▲ Clear and valid point

▲ Evidence provided in relation to technology

▲ Further detail provided for this type of technology

▲ Connects with the question

▲ A valid point with knowledge about the categories for strategies (proactive, reactive)

Technology has developed in the last 20 years and there have been many advances in increasing knowledge about the causes and effects of geophysical hazards. Technology can use hardware and software to try and reduce the risk from geophysical hazards. Geophysical hazards can be volcanoes and earthquakes. This essay will discuss the different ways in which technology helps to increase resilience in places in order that people and property are able to survive when an event takes place.

When an earthquake strikes, the introduction of technology has helped people to react more quickly and be able to evacuate from an area at risk. In order to warn people about incoming tsunamis, NOAA has installed a system of 39 buoys across the Pacific Ocean which will transmit information when a tsunami is created. The information is transmitted to a Tsunami Warning Center, such as in Alaska, and alerts are sent out via the media. This enables people to evacuate an area as the warnings are issued at least 3 hours before the tsunami reaches land. This ensures that moveable items can be moved, therefore reducing the risk and increasing the resilience for a place, such as major cities along the west coast of the USA.

Another strategy that is more proactive than reactive (the tsunami alert is a reactive strategy) is constructing buildings that can withstand major seismic waves. There are different techniques for making buildings more resilient:

deep foundations, dampers, and a reinforced structure. An example of a building that has deep foundations is the Transamerica building in San Francisco, which is a city that is prone to major earthquakes. The building looks like a pyramid and its foundations are sunk 52 feet underground which prevents the building from collapsing. Dampers have been constructed in one of the world's tallest buildings, the Taipei 101 building in Taiwan, which can swing up to 5ft inside the building to counteract the swaying of the building during an earthquake. Finally, the Torre Mayor building in Mexico City is constructed with reinforced concrete and steel which provides a rigid framework for the building. These buildings have further construction techniques in their design since more than 1 technique is needed to ensure resilience. People living and working in these buildings and the surrounding area will feel more secure knowing that these buildings will not collapse and there is no need to evacuate them during an earthquake.

When a volcanic eruption takes place, technology is used via the media to alert people that live near the volcano about the threat from ash or lava. This is something that happens in countries at different levels of development and it was used when Mount Kileaua in Hawaii erupted in 2018 when television was used to warn people.

Overall, technology has been used in order that places are able to become more resilient prior to and during a geophysical event. These measures have been implemented in places at different levels of development.

▲ Knowledge about resilient construction

▲ Three appropriate examples with accurate information

▲ Links to the question

▲ Appropriate example

▼ Limited detail for this example and the point being made is quite simplistic

This response is much stronger for earthquakes than volcanoes, and there is a lack of balance. Examples are included throughout with some detail provided although more detail would ensure that the response obtained a higher mark.

Marks 7/10

E LEISURE, TOURISM AND SPORT

This optional unit examines the relationship between increasing economic development and the evolving tastes for leisure and touristic activities. Places are examined in terms of their physical and human attributes, and also how governments can utilize these attributes as a route for development. The unit assesses the role of corporations in tourism, and the impacts of a rural festival and a large-scale international sporting event.

You should be able to show:

✔ how human development **processes** give rise to leisure activities;

✔ how physical and human factors shape **places** into sites of leisure;

✔ the varying **power** of different countries to participate in global tourism and sport;

✔ future **possibilities** for management of, and participation in, tourism and sport at varying scales.

E.1 CHANGING LEISURE PATTERNS

• **Leisure** – time free from the demands of work when a person can enjoy hobbies or sports.

• **Tourism** – when people travel to a place that is outside their home environment for no more than one year in duration for reasons such as leisure and business.

You should be able to show how human development processes give rise to leisure activities:

✔ The growth and changing purpose of leisure time for societies in different geographic and developmental contexts;

✔ The categorization of touristic activities (cost, duration, destination) and sporting activities (cost, popularity, site);

✔ The link between economic development and participation in leisure activities;

    ✔ Detailed examples to illustrate recent changes in participation for two or more societies at contrasting stages of development;

✔ Factors affecting personal participation in sports and tourism, including affluence, gender, stage in lifecycle, personality, place of residence.

The growth and changing purpose of leisure time for societies in different geographic and developmental contexts

A two-day weekend was introduced in China in 1995—previously, people only had one day off a week. In 2008, this changed again and a paid holiday system was introduced. The amount of paid holiday time that employers legally provide in China is one of the lowest in the world compared to countries with a similar or higher level of economic development.

In the UK, a country with a GDP per capita four times larger than that of China, workers receive at least 28 days paid annual leave per year if they work five days per week.

Content link

The growth of megacities in China is explored in unit 1.2, and the resulting social and environmental stresses are looked at in option G.3.

Leisure activities have evolved in China. As cities have grown into megacities, there has been an increase in a range of social and environmental stresses and as a result of this the outdoor leisure industry has grown as people try to escape urban areas.

The growth of technology has meant that young people now spend several hours each week participating in online gaming. The growth of the middle class in China means that people have the money to travel outside of the cities and to purchase technology. Some traditional leisure activities remain popular—elderly people in China play group games such as Mahjong.

The categorization of touristic activities (cost, duration, destination) and sporting activities (cost, popularity, site)

The cost of touristic and sporting activities varies according to following:

- Mode of transport

- Accommodation

- Duration of stay

- Distance travelled from place of origin to the destination

- The activities undertaken at the destination

- The equipment needed for a sporting activity.

Some sports will be more popular than others, and factors such as historical connections to a sport could influence the number of organizations and players connected to a sport in a country, for example, cricket in Commonwealth countries.

The link between economic development and participation in leisure activities

As countries develop economically, it is likely that there will be greater participation in leisure activities due to an increase in paid holiday time and a greater amount of disposable income.

There are a range of other economic factors such as a person's salary, the cost of living or the financial stability in a place, for example.

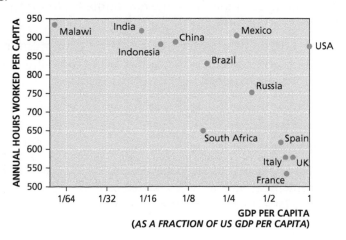

> **Concept link**
>
> **PROCESSES:** Socio-economic and political change bring societal developments which will increase access for people to participate in activities relating to leisure and tourism. Conversely, these dynamic processes may also reduce access for particular groups and produce differences within and between countries.

> **Content link**
>
> The growth of the "new global middle class" is examined in unit 3.1.

> **Test yourself**
>
> **E.1 Suggest** how economic factors may either increase or decrease a person's involvement in leisure activities. [3+3]
>
> **E.2 Describe** the relationship between GDP per person and number of hours worked as shown in figure E.1.1. [3]

> **>> Assessment tip**
>
> When describing information in a chart, it is sometimes necessary to describe how trends or relationships can experience subtle changes. The previous question demonstrates this, since while the relationship is generally negative, the negativity in the relationship is much clearer at higher levels of GDP.

◀ **Figure E.1.1.**
Hours worked vs GDP per capita (as a fraction of GDP per capita)

Source of data: American Economic Review 2016, 106(9): 2426–2457

Case study: Recent changes in participation in the UK

In recent years in the UK, the introduction of new technology has meant that people are now spending more leisure time at home. For example, a wider range of electronic devices and increased internet bandwidth have given rise to the popularity of movie streaming services. Healthy activities and gym membership have also increased as issues linked to unhealthy lifestyles have been highlighted by the media.

Factors affecting personal participation in sports and tourism, including affluence, gender, stage in lifecycle, personality, place of residence

As already discussed, affluence influences a person's involvement in leisure, and this also applies to a person's ability to engage in tourism. In the USA, children from low-income backgrounds are far less active than those from higher-income backgrounds according to the Aspen Institute's Sports and Society programme. Parks in low-income areas do not tend to have organized activities for children. Due to having less disposable income, less affluent young people are not able to engage with equipment-based sports such as golf.

A person's gender may influence which sports they participate in. Traditionally in some countries certain sports are played more often by one gender than the other.

In recent years, there have been some changes such that females can have access to the same sports as males. In July 2017, the government in Saudi Arabia removed a ban on females in public schools taking part in physical education (PE) lessons. It is hoped that this change will also allow females outside of school to take part in sporting activities during their leisure time.

Personality will influence a person's travel destination. For example, a person who is a risk-taker might be interested in visiting places in order to participate in rock climbing. A person's location may influence their ability to travel due to their access to transportation networks. In addition, a person's country of residence or nationality may affect their ability to travel to another country, due to restrictions on movement between the countries or visa restrictions for citizens from certain countries. For example, in 2018 the US government implemented restrictions on people travelling to the USA from countries such as Iran or Sudan.

In most countries young and elderly people represent the demographic sections of the population that have the most time for leisure and tourist activities.

Socio-economic and political **processes** are taking place in countries at different levels of development which affect the participation of people in leisure and tourism activities.

Test yourself

E.3 Outline two reasons why people's participation in leisure activities has increased in two societies at contrasting stages of development.   [2+2]

>> Assessment tip

An example should be included for each reason and for the short response questions, an example would help to ensure that the additional marks for extending a description or explanation are credited.

Content link

Policies devised to improve gender equality are outlined in unit 1.3.

Test yourself

E.4 Distinguish between leisure and tourism.   [2]

E.5 Examine three factors that determine levels of participation in sporting activities.   [3+3]

>> Assessment tip

A definition of tourism is given at the beginning of this chapter, but there are various different definitions. For example, having to stay overnight in a place is a definition from one source but it is not part of the definition from the United Nations World Tourism Organization.

>> Assessment tip

For the answer to E.5, it is possible to discuss participation at a local, national or an international scale. The first part of this unit enables you to have knowledge for all three scales. Therefore, a range of different factors can be explained.

E.2 TOURISM AND SPORT AT THE LOCAL AND NATIONAL SCALE

You should be able to show how physical and human factors shape places into sites of leisure:

✔ Human and physical factors explaining the growth of rural and urban tourism hotspots, including the role of primary and secondary touristic resources;

✔ Variations in sphere of influence for different kinds of sporting and touristic facility, including neighbourhood parks and gyms, city stadiums and national parks;

✔ Factors affecting the geography of a national sports league, including the location of its hierarchy of teams and the distribution of supporters;

    ✔ Case study of one national sports league;

✔ Large-scale sporting, musical, cultural or religious festivals as temporary sites of leisure and their associated costs and benefits;

    ✔ Case study of one festival in a rural location, its site factors and geographic impacts.

Human and physical factors explaining the growth of rural and urban tourism hotspots, including the role of primary and secondary touristic resources

The growth of rural tourism hotspots often initially involves physical factors, such as a site of outstanding natural beauty (primary resource) and then a human factor, such as the building of a road network (secondary resources) to enable people to visit the site. In urban areas, hotspots tend to focus on sites that have unique historical and cultural value (primary resource), such as the Eiffel Tower in Paris, and secondary touristic resources could be the tour guides at each site.

Variations in sphere of influence for different kinds of sporting and touristic facility, including neighbourhood parks and gyms, city stadiums and national parks

There is a wide range of sporting facilities and tourist attractions and each of them will have a sphere of influence (SOI), which is the area from which people will be drawn to that place. A sporting facility that is quite common in a place may have a small SOI (such as a gym) whereas a sporting stadium such as a velodrome will attract people from further afield. In general, the more unique and well-known a tourist attraction is, the further people will be prepared to travel. The following are factors that will determine the SOI of a place:

- The size of the facility
- The transportation links
- Advertising
- Networks (museums or art galleries could be part of a network allowing access to many different places).

- **Primary touristic resources** – the human and physical resources that attract people to visit a place. For example, the historical architecture (human) or the climate (physical) of a place. Note that not all human resources are developed for tourism purposes.

- **Secondary touristic resources** – the human resources that are developed for tourism such as the provision of accommodation or transportation infrastructure.

- **Tourism hotspot** – a place or an attraction that receives a high level of visitor interest.

- **Sphere of influence** – the area from which people are drawn in order to visit a sporting facility or touristic attraction.

- **Hierarchy** – a structure that, in this context, represents the ranking of sporting teams at a national scale.

Test yourself

E.6 "Most sporting facilities tend to be located near the centre of a city." **Discuss** this statement. [10]

›› Assessment tip

Question E.6 provides an opportunity to offer different perspectives depending on the sport and the city that is discussed. Responses will either agree or disagree. There are many possible approaches to this question and the candidate's argument and conclusion are likely to depend on the examples chosen for discussion.

Concept link

PLACES: Places that contain tourism and sporting facilities attract people from varying distances and this can be due to a number of physical and human factors. The amount of people visiting places or using facilities will subsequently alter the character of such places which may, in turn, influence the popularity among visitors and users, such as becoming less appealing due to overcrowding.

Factors affecting the geography of a national sports league, including the location of its hierarchy of teams and the distribution of supporters

Case study: The football league in England and Wales

Football is the most popular sporting pastime in England and Wales and there is a well-established league structure for professional and non-professional teams. In the National League System, there are 57 leagues and 84 divisions or tiers within leagues and the majority of players in these leagues will be amateurs or non-professionals. Above this level, there is a league called the English Football League which consists of League 2, League 1 and the Championship league, and teams from Wales also play in these leagues. The players in these teams will be professional. Above the Championship, there is the Premier League which includes many well-known football teams. Again, teams from Wales also play in this league.

Figure E.2.1 shows the location of the teams in the English Football League.

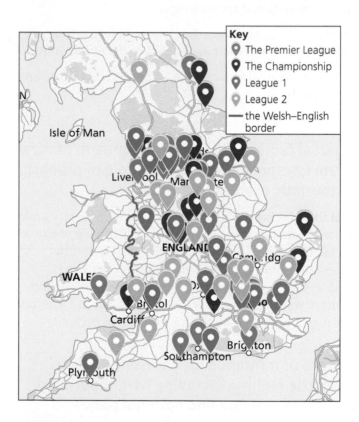

▶ **Figure E.2.1.** Location of teams in the English Football League and Premier League for the 2017/2018 football season

Test yourself

E.7 Describe the distribution of teams in the English Championship league. [3]

E.8 Explain why the map is not effective in showing the distribution of teams. [1+1]

E.9 Suggest how the data could be presented in a more effective way. [1+1]

The term "supporter" can refer to people actually travelling to the sporting venue to watch the team or alternatively watching the match on TV.

A number of factors affect a person's choice about which team they will support, such as the cost of travel or the link between the club and the local community, for example.

Test yourself

E.10 Referring to a national sports league you have studied, **suggest** the factors that have determined the locations of its teams.   [3+3]

Assessment tip

The example provided in this chapter is football, but a range of different sporting leagues can be explained for this question.

Large-scale sporting, musical, cultural or religious festivals as temporary sites of leisure and their associated costs and benefits

Festivals can be dedicated to different aspects of a location's culture, for example music, food, clothing and religion. Festivals can last for a short period such as an afternoon or for an extended weekend, and they can exist in both urban settings (for example, Notting Hill Carnival, London) and rural settings (for example, Burning Man Festival, Nevada).

As with any large-scale event, there are costs and benefits associated with festivals.

Case study: Bonnaroo Music and Arts Festival, USA

The Bonnaroo Music and Arts Festival is a four-day festival held in rural Tennessee, USA, 60 miles southeast of Nashville. It began in 2002 and attracts approximately 75,000 people when it is held each year in early June. The site is a 700-acre farm during the rest of the year. Festivalgoers reach the site via car or by using shuttle buses which run from the centre of Nashville or its airport, and hotels in the city run shuttle buses each day.

There are a number of geographic impacts associated with the festival:

- A road approaching the site is going to have to be upgraded since a bridge cannot handle the heavy traffic visiting the site. Local people are worried that this will encourage more traffic in the local area.

- Crimes have been committed, and people have been arrested each year, mainly for supplying drugs.

- The heat and humidity have caused some fatalities at the festival as well as illnesses such as heat exhaustion.

- The festival contributes approximately US $51 million to the local and state economies via the money spent by attendees and the tax revenue levied on the organizers.

- The festival reinforces the cultural history of music in this part of America with Nashville nearby.

- The Bonnaroo Works Fund was set up in 2009 which supports local and national organizations in the arts, education and environmental sustainability. About US$7 million had been donated to various organizations by 2017.

Test yourself

E.11 Identify the primary and secondary tourist resources for the Bonnaroo Music and Arts Festival.   [4]

E.3 TOURISM AND SPORT AT THE INTERNATIONAL SCALE

• **Niche tourism** – a specific tourism product that is tailored to meet the needs of a particular audience/market segment such as heritage tourism or movie location tourism.

• **TNC** – a transnational corporation or company that operates in a number of different countries.

You should be able to show the varying power of different countries to participate in global tourism and sport:

✔ Niche national tourism strategies with a global sphere of influence, including adventure tourism, movie location tourism and heritage tourism;

✔ The role of TNCs in expanding international tourism destinations, including the costs and benefits of TNC involvement for different stakeholders;

✔ Costs and benefits of tourism as a national development strategy, including economic and sociocultural effects;

✔ Political, economic and cultural factors affecting the hosting of international sporting events, including the Olympics and football World Cup events;

    ✔ Case study of costs and benefits for one country hosting an international event.

Concept link

POWER: Involvement in tourism can provide a pathway to economic development. Mainstream tourism products, as well as niche attractions, can attract a significant number of visitors and ensure the countries become more powerful economically. Any changes to global tourism markets can have a detrimental impact upon countries also, such as a global recession.

Test yourself

E.12 Outline what is meant by heritage tourism. [3]

E.13 State an example of heritage tourism. [1]

Niche national tourism strategies with a global sphere of influence, including adventure tourism, movie location tourism and heritage tourism

Niche tourism products are designed with a very specific market in mind. Due to this unique aspect, the sphere of influence for this product can extend outside of national boundaries. Adventure tourism products can offer a once-in-a-lifetime experience for some, for example, scaling Mount Everest.

Movie location tourism exists around the world as people venture to visit the place where their favourite movie was made or a particular scene was filmed. The greater the popularity of the film, the wider the SOI will be. For example, New York City has been home to many memorable films such as *King Kong* and *Superman*.

Heritage tourism can represent a wide range of historic and cultural aspects of a place. People will travel long distances to visit areas with significant buildings or where a well-known event took place, for example, the Forbidden City in Beijing.

The role of TNCs in expanding international tourism destinations, including the costs and benefits of TNC involvement for different stakeholders

A number of transnational corporations (TNCs) are involved in international tourism, and their presence in a country and destination can have a range of costs and benefits for different stakeholders.

Case study: Tourism in the Gambia

The Gambia in West Africa has a TNC presence in the country. It aims to boost its GDP via developing international tourism destinations along a 10-kilometre strip of land on the coast. Its economy is based on tourism, agriculture and remittances (the country experiences a large amount of out-migration, and remittances are the money that is sent back to the Gambia). Twenty large hotels accommodate 90% of all visitors, domestic or foreign.

The first TNC to arrive in the Gambia was Ving, a Norwegian company, in the 1960s. From that point on more hotels were built by TNCs arriving from other European countries.

The cost of air travel to Africa can often be expensive, so by creating affordable holiday deals that include air travel and accommodation, tour operators make it a lot easier for tourists to visit the Gambia, which in turn gives them significant power in controlling the flow of international visitors into the country.

The key benefit of TNC involvement is:

- Employment—42,000 people are involved in tourism-related activities; there are more jobs in foreign-owned hotels than domestic-owned hotels.

The costs of TNC involvement are:

- Jobs are low-skilled with a low salary and they are seasonal;

- Gender equality tends to be lower in foreign-owned hotels compared to domestic-owned hotels;

- Management or skilled positions tend to be held by Europeans rather than Gambians;

- The percentage of locally-sourced food is low in foreign-owned hotels, sometimes only 5%.

Costs and benefits of tourism as a national development strategy, including economic and sociocultural effects

The contribution that tourism makes to GDP on the continent of Africa increased from 6.8% in 1998 to 8.5% in 2017.

Economic effects include job creation and the sourcing of local food to support the agriculture industry. Countries will often introduce tourism taxes in order to help raise funds and reduce debt, for example, Barbados in 2018.

Sociocultural effects include the exchange of cultural traits between international visitors and locals thus furthering intercultural understanding. Countries such as Tanzania have improved gender equality by launching a range of female-driven tourism enterprises with women running and directing tour companies.

Test yourself

E.14 Using an example that you have studied, **explain** a strategy to develop tourism at a national scale. [2+2]

Political, economic and cultural factors affecting the hosting of international sporting events, including the Olympics and football World Cup events

There are several reasons why countries choose to put themselves forward to host international sporting events like the World Cup or the Olympics, such as the prestige of hosting an event and job creation. The creation of new sporting venues, adjoining facilities and infrastructure can be a strategy for regenerating urban areas.

Case study: Delhi Commonwealth Games, 2010

The Delhi Commonwealth Games in 2010 provided the following benefits:

- A new terminal was opened at the international airport.
- The Metro was expanded, with connections between the airport and city centre.
- 4,000 new buses were introduced that are powered by compressed natural gas (CNG).
- Roads were improved, and cycle paths were introduced.
- Air quality norms were introduced.
- Around 4,000 jobs were created.

The costs were as follows:

- 200,000 people were forcibly displaced to make way for infrastructure. Almost 8,000 flats were built in Bawana for some of those displaced, but many were reluctant to live there since there was no bus service to the city centre as travel time was 4 hours.
- Workers brought in from poor states such as Bihar did not receive the money they were promised and worked in unsafe conditions.
- The sports facilities were not used post-Games and have deteriorated because they are not generating enough revenue for their upkeep.
- Corruption was uncovered, with 16 projects containing irregularities, and some officials were taken to court.
- Taxes were increased in order to sustain the city's budget after spending on the Games. The original cost of the Games was estimated at US$270m but this increased to US$4.1 billion.

The success from growth in leisure and tourism is linked to the power of different stakeholders such as TNCs, national government and the citizens of a place.

>> **Assessment tip**

For questions about costs and benefits, the acronym SEEP will be useful: address the Social, Economic, Environmental and Political effects (this also applies to causes). Bear in mind that causes and effects occur at different scales. Also, benefits and problems can change over time, such as the short-term benefits versus the long-term costs of a sporting event.

Test yourself

E.15 Suggest reasons why a country's gross national income (GNI) increases before and during a major sporting event.   [3+3]

E.4 MANAGING TOURISM AND SPORT FOR THE FUTURE

You should be able to show examples of future possibilities for management of, and participation in, tourism and sport at varying scales:

✔ The consequences of unsustainable touristic growth in rural and urban tourism hotspots, including the concept of carrying capacity and possible management options to increase site resilience;

✔ The concept of sustainable tourism, including the growth of ecotourism;

  ✔ One case study of sustainable tourism in one low-income country;

✔ Factors influencing future international tourism, including greater use of social media, international security and diaspora growth;

✔ The growing importance of political and cultural influences on international sport participation, including international agreements, inclusion via changing gender roles and the growing importance of the Paralympics.

The consequences of unsustainable touristic growth in rural and urban tourism hotspots, including the concept of carrying capacity and possible management options to increase site resilience

Unsustainable touristic growth can have economic, social and environmental consequences for a destination. Excessive visitor numbers in hotspots can result in economic losses as a place gains a bad reputation and people do not want to go there. There can be social problems as local people begin to resent the overcrowding that occurs, and the environment can suffer due to noise and visual pollution.

Case study: Tourism in Barcelona

Barcelona has felt some of these impacts in recent years due to a growth in the number of tourists visiting the city, which is the 20th most visited city in the world.

Local people rent out homes and rooms via applications such as Airbnb, and people have been buying properties with the sole purpose of renting them out. This has created economic pressures on locals as rents have increased dramatically. The amount of Housing Used for Tourism (HUTs) grew from around 81 properties in 2005 to just under 10,000 in 2014. Anti-tourist graffiti has appeared throughout the city and a tourist bus was attacked in 2016.

In 2015, the Special Urban Plan for Tourist Accommodation Plan (PEUAT) was passed in order to manage tourism. Restrictions were introduced via HUT permits in four different zones:

• Zone 1: No more tourist rental licences or hotel licences will be granted.

Carrying capacity – the maximum number of people that may visit a tourist destination at the same time without causing destruction of the physical, economic or socio-cultural environment and an unacceptable decrease in the quality of visitors' satisfaction.

Resilience – the ability for a destination to overcome problems due to unsustainable growth in tourism numbers (and other factors such as security concerns).

Sustainable tourism – tourism that takes full account of its current and future economic, social and environmental impacts, and addresses the needs of visitors, the industry, the environment and host communities.

Diaspora – the dispersion of people from a country to a range of other countries.

Concept link

POSSIBILITIES: As places become more popular from tourism, and the possibilities for economic development become a reality, it is often necessary for alternative tourism products to be developed to ensure that there is economic, social and environmental sustainability.

- Zone 2: Situated next to Zone 1, this area has the same restrictions as Zone 1 in terms of new licences for accommodation, although if one establishment closes, a new licence may be issued.

- Zone 3: There is less impact from tourist accommodation in this zone, therefore 387 new licences will be granted when licences expire in other zones, and within certain parameters.

- Zone 4: These are areas where no tourist rentals are permitted.

▼ Figure E.4.1 Tourist market in Barcelona

The concept of sustainable tourism, including the growth of ecotourism

The term "sustainable tourism" is often intertwined with terms such as "ecotourism". However, there are distinctions between them that you need to be aware of. Ecotourism is concerned with the conservation and protection of biodiversity. Sustainable tourism is more comprehensive by ensuring not only environmental sustainability in a destination but also economic and sociocultural sustainability.

Case study: Sustainable tourism in Zimbabwe

Wilderness Safaris in Zimbabwe is a sustainable tourism company that has had a presence in the country since 1995. It provides safaris in national parks such as Hwange, which is home to the highest concentration of elephants in Africa. Revenue generated by the company supports eight schools in an area where the majority of families rely on subsistence agriculture. In addition, Wilderness has funded a women's cooperative that makes jewelry and a clinic that provides medical assistance to nine villages.

Wildlife is sustained via boreholes that are dug to provide water for elephants (surface water is limited) and Wilderness has covered the cost of at least 12 of them.

The company has also created a scheme called Children in the Wilderness, which runs weekly environmental groups in order to educate primary-school students about ecological value.

Some of the company's camps use only renewable energy which provides energy for lighting and for purifying water. Wilderness has also been involved in a campaign to prohibit the illegal hunting and tracking of the pangolin.

Factors influencing future international tourism, including greater use of social media, international security and diaspora growth

Technology is continually changing how people plan their international tourism experiences. Social media is advertising destinations via a range of different applications while reviews of primary and secondary touristic resources are updated on websites such as TripAdvisor.

In terms of security, the threat of conflict and terrorism is expected to impact tourism in the future. However, in some parts of the world there has been a reduction in conflict. Colombia, for example, is developing ecotourism in parts of the country that were previously controlled by the rebel group, FARC. On the other hand, the movement of travellers has been restricted in some countries, for example, as a result of the USA's ban on people travelling from certain countries such as Iran.

Diaspora growth has been identified as an area for increasing tourism revenue in some countries. For example, in 2017, Jamaica announced that between US$250 million and US$300 million is generated annually from Jamaicans returning to the island.

The growing importance of political and cultural influences on international sport participation, including international agreements, inclusion via changing gender roles and the growing importance of the Paralympics

Stereotypes and traditional roles have been altered in recent years by increased gender equality in sports and a greater understanding about the Paralympics. In terms of gender, the International Olympic Committee (IOC) has a Gender Equality Review Project alongside a Women in Sport Commission, both of which aim to address issues of gender inequality in sport. One objective is to ensure that female participation in the Olympic Games reaches 50% by 2020.

The Paralympics is gaining importance in society as more people visit the event. More people are also participating (there were 9.9% more female athletes in Rio 2016 compared to London four years earlier). Furthermore, viewing figures are increasing (for example, there were 1.84 billion viewers for Athens 2004, but 4.11 billion viewers for Rio 2016).

There are possibilities for increasing equality and economic, social and environmental sustainability in leisure and tourism, although without adequate management the positive impact may not be realized.

The table shows the world's 15 most visited cities in 2017.

▼ Table E.4.1. The 15 most visited cities in the world, 2017

City	Visitors (millions)	City	Visitors (millions)	City	Visitors (millions)
Bangkok	20.2	Tokyo	12.5	Istanbul	9.24
London	20	Seoul	12.44	Barcelona	8.9
Paris	16.1	New York City	12.4	Amsterdam	8.7
Dubai	16	Kuala Lumpur	12.1	Milan	8.4
Singapore	13.45	Hong Kong	9.25	Osaka	7.9

Source of data: Mastercard Global Destination Cities Index (2017)

a) i) Determine the range of values for visitor numbers in 2017. [1]

   ii) **State** the median value of visitor numbers in 2017. [1]

b) **Outline** one way in which high visitor numbers may impact negatively on local communities. [2]

c) Using an example, **explain** one strength and one weakness of ecotourism for local communities. [3 + 3]

Essay

Examine ways in which tourism can affect a country's development. [10]

How do I approach these questions?

a) i) A quick calculation is required for this question and it is recommended that you show your working. You should also include the unit of measurement (in this case, millions of visitors).

   ii) This is quick question worth only 1 mark. No sentence is required, simply state the median value (remember that the median is the middle number).

b) This question is worth 2 marks, so be prepared to be quite specific about the negative impact, or alternatively include an example. Ideally, you should include both.

c) You should structure your answer to this question as two separate paragraphs, for the strength and weakness respectively. These two points will require development, and you should refer to an example in your answer.

Essay

This is quite an "open" question in which you can discuss a range of effects due to the development of tourism such as economic, social and environmental consequences. These consequences could come from different tourism sectors, such as mass tourism or sustainable tourism, and there are both costs and benefits from tourism. Remember that the command term is "examine", so an assumption that tourism can only be beneficial (or the opposite) must be challenged in your response. Key concepts such as **place** could be discussed as well as **scale** since you will have studied tourism as a national development strategy. With this in mind, you could discuss the national benefits from such a strategy, such as the contribution to GDP, as well as the local disadvantages, such as locals only being employed in unskilled jobs.

You could also take a temporal approach by examining how tourism can affect a place over time. You may have studied the Butler model, which demonstrates how costs and benefits can occur as a destination develops and how these perspectives may differ depending on the stakeholder involved (for example, a local environment group versus a hotel owner).

a) i) 20.2 – 7.9 = 12.3 million visitors.

Mark 1/1

ii) 20.2 – 7.9 = 12.3 million visitors.

Incorrect, since the range between the 1st and 15th value has been stated again. New York City with 12.4 million is the median value (8th consecutive value out of 15).

Marks 0/1

b) High numbers of visitors can negatively impact on local communities due to the congestion and overcrowding of tourists on a city's streets. This is seen in cities such as Barcelona and Vienna and it means that it is more arduous and time consuming for people to travel to places within the city for their jobs or going shopping.

▲ Negative impact given

▲ Example given and point developed further

Marks 2/2

c) One strength for local communities is that their environment is not being damaged by the tourists, thus they are sustainable for the future. This has been seen in the Monteverde cloud forest in Costa Rica.
A negative impact from ecotourism is that there is a lack of more widespread economic growth in an area since visitor numbers are minimized in order to protect the environment. The development of mass-market tourism will create a larger amount of jobs for the local community and these will in turn create many indirect jobs.

▼ There is a lack of development for this valid point (although a relevant example is included)

▼ A valid reason is given. Further development needed to get the third mark

Marks 4/6

Essay

Examine ways in which tourism can affect a country's development.

The term 'development' is often used to describe a country's economic development although it can also be applied to the country's social development, such as health and education. Tourism is when people decide to visit a destination away from their normal home but not for longer than 1 year. Some countries

▲ Knowledge is shown here with a valid definition

▲ Another valid definition

▲ Appropriate terminology

▲ Provides a clear thesis statement which provides a foundation for evaluation

▼ Could provide an outline of the examples to be discussed which would improve the structure

▲ Clear point

▲ Provides evidence of tourism growth which applies to the question

▲ Relevant example with some detail

▲ A final sentence that connects with the thesis statement and question

▼ No point made at the beginning of this paragraph which would help the structure

▲ Example of economic benefits with explanation

▲ Example and explanation of social benefits

with unique primary touristic resources such as a beach or a historical building have the resources available to be able to increase their GDP with an influx of tourists. Alternatively, an influx of tourists can exceed a place's carrying capacity. Therefore, I agree and disagree with the statement since there are costs as well as benefits.

Greece has seen tremendous benefits from tourism during its modern history. Mass tourism has developed in the country and it is a very popular destination for many Europeans. In 1950, 33,000 tourists visited Greece which increased to 27 million by 2017. Greece suffered an economic collapse in 2008 and suffered a recession for 6 years. It was helped with assistance from the IMF and other European countries but tourism helped it recover and it contributed approximately 25% of Greece's GDP with almost 30% of all employment connected to tourism. Therefore Greece is an example of a country that experienced economic development from tourism, suffered a recession but then used tourism once more to further develop economically.

The Maldives, which has a GDP per capita figure almost half that of Greece and it receives almost 75% of its GDP from tourism and 16% of all jobs are supported by tourism. It is important to recognize that there are indirect jobs supported by tourists as well as since tourism workers will spend their money in local restaurants and supermarkets which means that other employment is supported. All of these jobs will be taxed and will contribute money to the government which can then be spent on improving the social development of the Maldives such as health and education.

There can be a cost though as a destination can experience significant social problems whilst earning increased revenue from tourism and Ibiza is a great example for this. However, there are costs such as overcrowded beaches, traffic congestion and an increase in crime during the tourism season. In addition, Ibiza is famous for its clubs and the illegal drugs trade has increased over the last 30 years. Economically, there have also been problems since local people are renting out rooms and apartments (this was banned in 2018 though) so rents increased for local people who were looking for somewhere to live. Therefore social and economic problems can be the result of tourism development. This essay has shown the costs and benefits from tourism by examining tourism growth in Greece, the Maldives and Ibiza. This essay only somewhat agrees with the statement.

▲ Discussion of costs as well as benefits—therefore the answer is an evaluation

▲ An appropriate example

▲ Evidence for social costs

▲ Evidence for economic costs

▲ Connects with the thesis statement

A very short conclusion is present, which could be more detailed by reflecting on the key points of the essay. The response demonstrates that different places have varying experiences from tourism and examples from a range of countries at different levels of development show the student's breadth of knowledge.

Further knowledge could be included about the growth of tourism in each destination but the presence of evaluation ensures that the response will be credited with a mark from 7 upwards. Further explanation and some minor improvements to the structure of the essay would increase the mark awarded and the carry capacity of places could be explicitly mentioned in the main body of the essay.

Finally, more connections with the key concepts, such as the power of different stakeholders within the development and evolution of a tourism strategy, would improve the response.

Marks 8/10

F FOOD AND HEALTH

This theme examines the geography of food and health. Economic development is often accompanied by changes in diet and in disease pattern. However, neither food intake nor health is easy to measure. Food and health are closely related. The provision of food and health are influenced by gender, TNCs and governments.

You should be able to show:

✔ ways of measuring disparities in food and health between **places;**

✔ how physical and human **processes** lead to changes in food production and consumption, and incidence and spread of disease;

✔ the **power** of different stakeholders in relation to influence over diets and health;

✔ future **possibilities** for sustainable agriculture and improved heath.

F.1 MEASURING FOOD AND HEALTH

- **Food security** – food security for a population exists when all its people, at all times, have access to sufficient, safe and nutritious food to meet their dietary needs and food preferences for an active and healthy life.

- **Nutrition transition** – the change in diet that is associated with a population becoming wealthier (shifting from low income to middle income) and consuming more meat and dairy products.

- **Epidemiology** – the study of diseases.

- **Epidemiological transition** – the shift in the major diseases experienced as a population moves from being poorer to wealthier. For example, a decrease in infectious diseases but an increase in degenerative diseases.

You should be able to show ways of measuring disparities in food and health between places:

✔ Global patterns in food/nutrition indicators, including the food security index, the hunger index, calories per person/capita, indicators of malnutrition;

✔ The nutrition transition, and associated regional variations of food consumption and nutrition choices;

✔ Global patterns in health indicators, including health-adjusted life expectancy (HALE), infant mortality, maternal mortality, access to sanitation and the ratio between doctors/physicians and people;

✔ The epidemiological transition, the diseases continuum (diseases of poverty to diseases of affluence), and the implications of a global aging population for disease burden.

Global patterns in food/nutrition indicators

There are many inequalities in access to food and nutrition. Some of the data uses terms that are quite subjective, such as the Global Hunger Index, although they are comprised of many elements related to malnutrition and mortality.

The Food Security Index measures the affordability, availability and quality of food.

The Global Hunger Index (GHI) is a composite indicator, consisting of three main components (but four indicators). These include child mortality (as measured by the under-5 mortality rate), child undernutrition (stunting and wasting) and inadequate access to food.

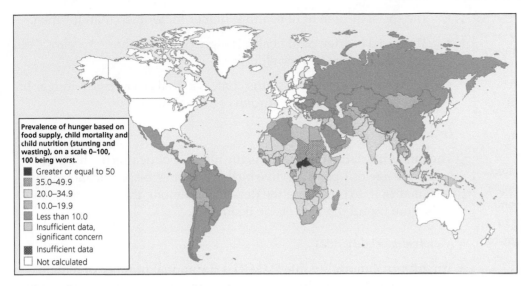

▲ **Figure F.1.1.** GHI, by severity, 2017

Access to food can also be measured by the number of calories per person per day, or by indicators of malnutrition such as weight/age or and height/age compared with national statistics.

The nutrition transition, and associated regional variations of food consumption and nutrition choices

The dietary changes that characterize the "nutrition transition" include qualitative and quantitative changes in diet. There is a shift towards a higher energy density diet with increased fat and added sugar, greater saturated fat intake (mainly from animal sources), and reduced intake from carbohydrates, dietary fibre, fruit and vegetables.

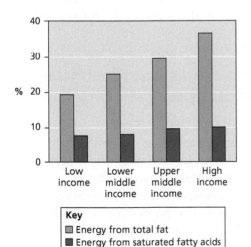

◀ **Figure F.1.2.** Percentage of total energy from fat and saturated fatty acids

Source of data: World Health Organization

Global patterns in health indicators

Health-adjusted life expectancy (HALE)

HALE is an indicator of the overall health of a population. It combines measures of both age- and sex-specific health data, and age- and sex-specific mortality data, into a single statistic. HALE indicates the number of expected years of life equivalent to years lived in full health, based on the average experience in a population. Thus, HALE is a measure of quantity of life *and* of quality of life.

Test yourself

F.3 Explain why the infant mortality rate is a useful indicator of development.   [2]

F.4 Suggest why the number of doctors per person is not a reliable measure of the quality of health care systems.   [2]

 Content link

The challenges associated with aging populations are looked at in unit 1.3.

- The social burden of ill health is higher for women than for men.
- The social burden of ill health is highest among those in "early" old age, not among the very elderly.
- Higher socio-economic status confers a dual advantage—longer life expectancy and a lower burden of ill health.

Infant mortality rate

The infant mortality rate (IMR) is the number of deaths in children under the age of 1 per 1,000 live births. It is an age-specific mortality rate, that is, it is comparing the death rates among the same ages, and so is more useful than the crude death rate.

Maternal mortality rates

The maternal mortality rate (MMR) is the annual number of female deaths per 100,000 live births from any cause related to or aggravated by pregnancy or its management. In 2016, South Sudan had the highest MMR with over 2,000 deaths per 100,000 live births.

Access to sanitation

Due to a combination of population growth and slow progress with the provision of facilities, the number of people in sub-Saharan Africa without access to sanitation has increased since 1990. In addition, there are rural and urban disparities: over 80% of the urban population has access to improved sanitation facilities compared to 51% in rural areas.

Global variations in access to doctors/physicians

Access to doctors varies from one doctor per 100,000 people in Burundi and one doctor per 50,000 people in Mozambique to one doctor per 280 people in Hungary and Iceland. There is also a disparity in the facilities available in hospitals and clinics.

The epidemiological transition, the diseases continuum and the implications of a global aging population for disease burden

The epidemiological transition refers to the shift in mortality from largely infectious, communicable diseases to those that are largely degenerative and non-communicable. In the last two centuries, there has been a decline in infectious diseases in many of today's HICs and a rise in degenerative diseases. The proportion of deaths due to cardiovascular disease (CVD) increased to between 10% and 35%.

Aging and the disease burden

Of the total global burden of disease, 23% is attributable to disorders in people aged 60 years and older. This accounts for about 50% of the health burden in high-income countries and 20% of the burden in low- and middle-income countries. The leading contributors to disease burden in older people are CVD (over 30% in people aged 60 years and over) and cancer (15%).

F.2 FOOD SYSTEMS AND SPREAD OF DISEASES

You should be able to show how physical and human processes lead to changes in food production and consumption, and incidence and spread of disease:

✔ The merits of a systems approach (inputs, stores, transfers, outputs) to compare energy efficiency and water footprints in food production, and relative sustainability in different places;

✔ The physical and human processes that can lead to variations in food consumption;

✔ The importance of diffusion (including adoption/acquisition, expansion, relocation) in the spread of agricultural innovations, and also in the spread of diseases, and the role of geographic factors (including physical, economic and political barriers) in the rate of diffusion;

✔ Geographic factors contributing to the incidence, diffusion and impacts (demographic and socio-economic) of vector-borne and water-borne diseases;

   ✔ One detailed example of a vector-borne disease and one detailed example of a water-borne disease.

> • **Systems** – a simplified way of looking at a feature (such as farming), by breaking it down into inputs (factors), processes and products.
>
> • **Water footprints** – a measure of the use of water by humans or nations and/or the amount needed to grow or manufacture products such as meat.
>
> • **Diffusion** – the way in which a feature (such as a disease) spreads.

The merits of a systems approach to compare energy efficiency and water footprints in food production, and relative sustainability in different places

A system is a simplified way of looking at a complex feature, by breaking it down into inputs (factors that affect the system), their relative size (stores), processes (the activities that lead to the production of food) and outputs (products of the system).

Energy efficiency ratios

The energy efficiency ratio is a measure of the amount of energy inputs into a system compared with the outputs. In a traditional agroforestry system, the inputs are very low but the outputs are relatively high. This contrasts with intensive farming where the inputs may be quite high but the returns may be relatively low.

Water footprints in food production

The projected increase in the production and consumption of animal products is likely to put further pressure on the world's freshwater resources. Large-scale commercial farming systems tend to have higher water footprints than small-scale subsistence farming systems.

The water footprint of meat from beef cattle (15,400 m³ per tonne) is much larger than the footprint of meat from sheep (10,400 m³ per tonne) or chicken (4,300 m³ per tonne).

Per tonne of product, animal products generally have a larger water footprint than crops.

Global animal production requires about 2,422 billion cubic metres of water per year. Most of this water (98%) is used in the production of feed for the animals.

▼ Table F.2.1. Energy efficiency ratios for selected farming systems (input:output of energy input)

Agroforestry	1:65
Hunter-gatherers	1:7.8
Cereal farm	1:1.9
Dairy farm	1:0.38
Greenhouse lettuces	1:0.002

Source of data: Adapted from Tivy, J., *Agricultural ecology*, 1990, Longman

Test yourself

F.5 Suggest why the energy efficiency ratio for agroforestry is higher than for dairy farms.   [2]

F.6 Explain why the water footprint for animals is higher than for crops.   [2]

Content link
The effect of income on the consumption of food and other resources is looked at in unit 3.1.

The physical and human processes that can lead to variations in food consumption

Income and level of education influence food choice. Diet may vary depending on the availability of income to purchase more healthy food. For a low-income family, price plays a larger role than taste and quality in deciding whether the food will be purchased. The variety of foods carried in neighbourhood shops may also influence diet.

The importance of diffusion in the spread of agricultural innovations, the spread of diseases, and the role of geographic factors in the rate of diffusion

Diffusion of innovations

The introduction of a new agricultural technique depends upon a number of factors including information regarding innovations, financial security, the personality of the adopter and the proximity to other adopters. Initially very few people adopt an innovation. As information becomes more widespread, and often the cost is reduced, increasingly more people adopt the innovation (figure F.2.1). However, some people are reluctant to change and will take a long time, if at all, to accept the new technique.

Assessment tip

Be sure to study diagrams carefully, and work out what they show before you begin an answer. In figure F.2.1 the blue line shows the number of adopters at any given time whereas the red line refers to the cumulative number of people that have adopted an innovation over time.

▼ **Figure F.2.1.** The diffusion of innovations

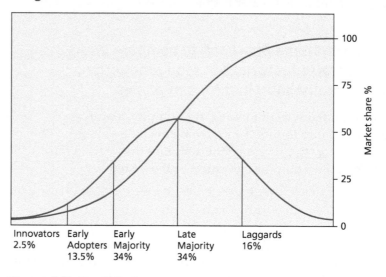

Test yourself

F.7 Suggest reasons why some people are reluctant to adopt an innovation.   [2]

Types of disease diffusion

There are several types of disease diffusion:

- **Expansion diffusion** occurs when the expanding disease has a source and diffuses outwards into new areas.

- **Relocation diffusion** occurs when the spreading disease moves into new areas, leaving behind its origin or the source of the disease.

- **Contagious diffusion** is the spread of an infectious disease through the direct contact of individuals with those infected.

- **Hierarchical diffusion** occurs when a phenomenon spreads through an ordered sequence of classes or places.

- **Network diffusion** occurs when a disease spreads via transportation or social networks.

Geographic factors contributing to the incidence, diffusion and impacts (demographic and socio-economic) of vector-borne and water-borne diseases

A number of geographic factors lead to the incidence, spread and impacts of diseases such as cholera (a water-borne disease) and malaria (an insect/a vector-borne disease).

Water-borne disease: Cholera

Each year there are between 1.3 million and 4.0 million cases of cholera, and 21,000–143,000 deaths worldwide. Most of those infected can be successfully treated with oral rehydration solution.

Cholera is an acute diarrheal infection caused by ingestion of food or water contaminated with the bacterium *Vibrio cholerae*.

Cholera transmission is closely linked to inadequate access to clean water and sanitation facilities. Typical at-risk areas include slums and refugee camps, where minimum requirements for clean water and sanitation are not being met.

The long-term solution for cholera control lies in economic development and universal access to safe drinking water and adequate sanitation.

Actions targeting environmental conditions include the implementation of adapted long-term sustainable water, sanitation and hygiene (WASH) solutions to ensure the use of safe water, basic sanitation and good hygiene practices to populations most at risk of cholera.

Vector-borne disease: Malaria

Malaria is a life-threatening disease caused by parasites that are transmitted to people through the bites of infected female *Anopheles* mosquitoes. It is preventable and curable.

In 2016 there were an estimated 216 million cases of malaria and 445,000 deaths. Africa accounted for 90% of malaria cases deaths (40% in Nigeria and DR Congo).

Total funding for malaria control and elimination reached an estimated US$2.7 billion in 2016. Vector control is the main way to prevent and reduce malaria transmission.

▼ **Figure F.2.2.** The global distribution of malaria

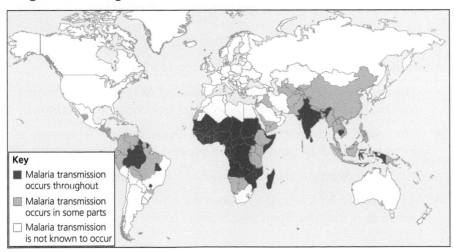

Key
■ Malaria transmission occurs throughout
▨ Malaria transmission occurs in some parts
□ Malaria transmission is not known to occur

Source of data: Centers for Disease Control and Prevention

Concept link

PROCESSES: Changes to the processes of food production and consumption take place at a range of different scales. Production, for example, can be affected by the access to shipping routes during geopolitical tension, or the availability of new technology that increases agricultural yields. These processes operate within a system, and the water-food-energy nexus is part of this system. The spread of different forms of disease is possible due to geographic processes such as the movement of people, or the lack of vaccines in a place.

Two forms of vector control—insecticide-treated mosquito nets and indoor residual spraying (IRS)—are effective in a wide range of circumstances.

Indoor residual spraying with insecticides is a powerful way to rapidly reduce malaria transmission. Its potential is realized when at least 80% of houses in targeted areas are sprayed.

Test yourself

F.8 Outline the ways in which diseasess may be spread.   [5]

F.9 (a) Describe the distribution of malaria as shown in figure F.2.2.   [3]

(b) Explain two reasons for the distribution that you have described. [2+2]

F.10 Compare the main conditions required for the development of cholera with the main conditions required for the development of malaria.   [2+2]

F.3 STAKEHOLDERS IN FOOD AND HEALTH

• **Famine** – the UN definition of a famine states that 20% of the population must have fewer than 2,100 kilocalories of food available per day; more than 30% of children must be acutely malnourished; and two deaths per day in every 10,000 people (or four deaths per day in every 10,000 children) must be being caused by lack of food.

You should be able to show the power of different stakeholders in relation to influence over diets and health:

✔ The roles of international organizations (such as the World Food Programme, Food and Agriculture Organization of the United Nations and World Health Organization), governments and NGOs in combating food insecurity and disease;

✔ The influence of TNCs (agribusinesses and the media) in shaping food consumption habits;

✔ Gender roles related to food and health, including food production/acquisition and disparities in health;

✔ Factors affecting the severity of famine, including governance, the power of the media and access to international aid;

✔ One case study of the issues affecting a famine-stricken country or area.

>> **Assessment tip**

Be careful about using the term "famine". It has a very precise meaning. Many places lack food security but do not have a famine.

The roles of international organizations, governments and NGOs in combating food insecurity and disease

Combating food insecurity

Many stakeholders contribute to achieving food security:

• The Food and Agriculture Organization (FAO), whose main aims include the eradication of hunger, food insecurity and malnutrition;

• The World Food Programme (WFP) aims to end global hunger. It focuses on food assistance for the poorest and most vulnerable;

• National governments may subsidize food production;

• A number of NGOs help to deliver food to those with insufficient access to it, such as Operation Hunger in South Africa.

Combating disease

The World Health Organization (WHO) is the part of the UN that deals with health issues.

There are other initiatives. For example, the Global Fund to Fight AIDS, Tuberculosis and Malaria was launched in 2002. The Global Alliance for Vaccines and Immunizations (GAVI) provides international financing for immunization coverage.

Médecins Sans Frontières (MSF—Doctors Without Borders) is an NGO that was founded in 1971 to provide emergency medical aid and a new brand of humanitarianism independent of governments. MSF, which was awarded the Nobel Peace Prize in 1999, is a worldwide movement owned and run by its staff. In 2015 it provided medical aid in over 70 countries.

- MSF treated 5,883 people for diphtheria in the Cox's Bazar district (in Bangladesh) between September 2017 and April 2018, most of them aged between 5 and 14.

- They also saw 4,680 cases of measles in that period.

- They treated 377 survivors of sexual and gender-based violence (SGBV) between August and April 2018. However, MSF believe they only treat a fraction of all cases.

The influence of TNCs (agribusinesses and the media) in shaping food consumption habits

The nutrition transition in LICs leads to a change in diet away from starchy staples to include more fruit and vegetables, meat and dairy, but there is also a tendency towards an increase in intake of processed, energy-dense, non-traditional foods which are often high in sugar, salt and harmful fatty acids, and are poor in micronutrients.

Multinational retailers have followed multinational food manufacturers, soft-drink companies and fast-food chains into food and drink sectors in virtually all countries; they have introduced the types of supply-chain management previously seen only in the developed world.

The move towards more Western-style diets in MICs and LICs may be seen as demand driven. Growing incomes, urbanization and female labour-force participation have led to a demand for convenience processed and fast food and eating out.

There has been a rapid expansion of supermarkets in MICs and LICs. In Latin America, supermarkets deliver 50%–60% of retail food sales. Modern food systems reduce the price of processed convenience foods relative to traditional staples and fresh fruit and vegetables. Modern manufacturers, fast-food and soft drink firms and supermarkets employ sophisticated marketing, often targeted at children, to encourage a preference for Western foods.

There is some evidence that supermarkets (and convenience stores) have reduced the prices of packaged foods relative to fresh produce, particularly in the early stages of supermarket penetration in a country. In Brazil, some packaged food was 40% cheaper than in traditional outlets.

Concept link

POWER: When there are issues connected with the healthcare and diets of people, it normally involves a significant section of the global population, covering an area of a country, continent or a socio-economic/demographic group within society. To tackle these problems, it can often involve a range of different stakeholders, each with different levels of power, but each with a vested interest in safeguarding the lives of those that struggle to get access to good and to combat disease.

Test yourself

F.11 Explain how TNCs influence global food consumption.　　[4]

Gender roles related to food and health

Gender, food security and nutrition

▲ **Figure F.3.1.** A woman prepares injera bread in Chencha, Ethiopia

In low-income countries, such as Ethiopia and Eritrea, rural women and men play different roles in guaranteeing food security in their households and communities. While men grow mainly field crops, women are usually responsible for growing and preparing most of the food and rearing small livestock, which provides protein. Rural women also carry out most food processing, which ensures a diverse diet, minimizes losses and can provide marketable products. Women represent about half of the food-producing workforce in South-East Asia and sub-Saharan Africa, but often as unpaid workers involved in subsistence farming.

🔗 **Content link**

Unit 5.1 examines ways of promoting gender equality in the workplace.

Gender and health

Life expectancy for women is generally higher than for men. This may be partly because men are more likely to have a more "self-destructive" lifestyle than women. However, more men work full-time than women, and retire at a later age, and that may hasten their death. Nevertheless, women in LICs have very physical jobs, which may explain, in part, the low life expectancy compared with HICs. Poverty and diseases are also likely to play an important part.

▼ **Table F.3.1.** Highest and lowest life expectancies: male and female (2015–20)

Rank	Highest female life expectancy	Years	Rank	Highest male life expectancy	Years
1	Monaco	93.6	1	Monaco	85.6
2	Hong Kong	87.4	2=	Hong Kong	81.7
3	Japan	87.3	2=	Iceland	81.7
4	Singapore	86.7	4	Switzerland	81.6
5	Italy	86.0	5	Israel, Italy	81.3
	Lowest female life expectancy			**Lowest male life expectancy**	
1	Swaziland	47.7	1	Swaziland	49.5
2	Lesotho	50.2	2	Lesotho	50.3
3	Sierra Leone	52.7	3	Cen. African Rep.	51.1
4	Chad	53.6	4	Chad	51.4
5	Cote d'Ivoire	53.8	5	Sierra Leone	51.5

Source of data: The Economist, *Pocket world in figures* (2017)

Test yourself

F.12 Study table F.3.1.
(a) Describe the main differences in life expectancy for the countries with the highest and lowest life expectancies for females and males.   [2+2]

(b) Suggest reasons for the differences that you have identified.   [3]

Factors affecting the severity of famine

There are many causes of famine. Prolonged low and/or unreliable rainfall may lead to water shortages and food shortages. Deforestation or overgrazing may cause soil degradation. Increased population pressure or a lack of secure land tenure could lead to a reduction in the amount of land per person. Decreasing affordability of food or a decrease in food entitlement (such as unemployment) could lead to outbreaks of famine. A lack of proper storage facilities may lead to increased food waste. However, the main factor is likely to be political —civil war disrupts farming, transport and access to aid. Civil war was the main factor in causing a famine in South Sudan in recent years.

Case study: Famine in Africa and the Middle East, 2017

In 2017, South Sudan was declared to be in a state of famine by the UN. It was the first time since 2011 that the UN had used the term. A further

1.1 million people were said to be in an "emergency" situation. Some 250,000 children under the age of 5 suffer from "severe acute" malnutrition. Nearly 6 million people relied on food aid during 2018.

Three other countries—Nigeria, Somalia and Yemen—were said to be at a "credible risk of famine". Between the four countries, 20 million people are at risk of starvation. The factor that they share is war. Since 2013, over 25% of South Sudan's population have fled their homes to escape ethnic killings. People who flee cannot harvest their crops or work to pay for food.

Test yourself

F.13 Define the term famine. [2]

F.4 FUTURE HEALTH AND FOOD SECURITY AND SUSTAINABILITY

You should be able to show examples of future possibilities for sustainable agriculture and improved health:

✔ Possible solutions to food insecurity, including waste reduction;

  ✔ One case study of attempts to tackle food insecurity;

✔ Advantages and disadvantages of contemporary approaches to food production, including genetically modified organisms (GMOs), vertical farming and in vitro meat;

✔ The merits of prevention and treatment in managing disease, including social marginalization issues, government priorities, means of infection and scientific intervention;

✔ Managing pandemics, including the epidemiology of the disease, prior local and global awareness, international action and the role of media;

  ✔ One case study of a contemporary pandemic and the lessons learned for pandemic management in the future.

- **Epidemic** – a fast-spreading outbreak of a disease.
- **Pandemic** – a global epidemic.
- **In vitro meat** – Cultured or synthetic meat produced in a laboratory from stem cells rather than from an animal that has ever lived.

Possible solutions to food insecurity

Case study: Achieving food security in Bangladesh

Food insecurity in Bangladesh is affected by international trade, land scarcity, the need to increase production of nutritional food, natural hazards and climate change. Food security remains an issue at national, household and individual levels. Bangladesh has made significant progress in improving food security by increasing production of rice using irrigation and high-yielding varieties. Increased emphasis on rice has necessitated increased imports of other foods. The government has also invested in storage facilities for rice, and cold-storage facilities for meat, fish, eggs and potatoes. The transport infrastructure has been upgraded to enable faster and better distribution of food, including imports.

>> **Assessment tip**

Despite progress being made to find solutions to food security, not everyone will benefit. It is good practice to identify those who will not benefit from improved food security, as well as those who will.

Advantages and disadvantages of contemporary approaches to food production

The advantages and disadvantages of genetically modified organisms (GMOs)

There are many advantages related to GMOs. For example, food supplies become more predictable, and food quality can be improved by the

Test yourself

F.14 Explain how it is possible to achieve food security. [4]

POSSIBILITIES: Environmental and social sustainability refers to the improvements to the natural landscape such that future generations have access to the similar or improved levels of nutrition as the current generation. With rising populations and increased consumerism, plus the impact of global climate change, innovative food production techniques are needed to increase production such that providing adequate nutrition is possible. This also requires efficient distribution strategies, and thus there are a range of different stakeholders involved in the food production process.

introduction of more vitamins. GMOs can be modified to last longer and may even have medical benefits (increased nutrients, proteins and vaccines designed into the food). There may be less need to use herbicides and pesticides, as genetic resistance is designed into the GMO.

However, GMOs may cause antibiotic resistance. Crops that are genetically modified may produce seeds that are genetically modified. It is possible for genes to get into wild species—a number of weed species are known to be resistant to the herbicide atrazine. GMOs are heavily controlled by TNCs, and independent research regarding their impacts is generally not allowed.

Vertical farming

Vertical farming allows crops to be grown throughout the year, and by day and night, as it uses LED lighting. It reduces transport costs as many vertical farms are found in high-rise buildings in cities. This reduces air pollution and emissions of CO_2. It uses minimal water, as water use can be controlled and recycled. Vertical farming also grows food organically—no pesticides are needed as there are no pests to damage the crops.

However, vertical farming could lead to a loss of jobs in the transport sector. The cost of pollination increases as there are no insects to pollinate the plants naturally. This increases the costs of production. There is a great reliance on technology, for lighting, heating and irrigation. This means that if there are power cuts, the crops could die. Vertical farming occurs mainly in HICs in urban areas, for example "Plenty" in San Francisco.

In vitro meat

In vitro meat (IVM) already exists but is very expensive. It refers to meat that is grown in a laboratory rather than on a farm. IVM offers a potentially more environmentally friendly and animal-welfare friendly, ethical and disease-free type of farming compared to conventional meat production systems that are energy- and water-intensive and contribute to local and global pollution. On the other hand, IVM is perceived as unnatural, potentially less tasty and likely to put farmers out of business.

The merits of prevention and treatment in managing disease

Preventative treatment means adopting policies and lifestyles that will reduce the risk of disease. This may range from people having a healthy diet to not smoking or drinking to excess to reduce the risk of cancer, heart attacks and strokes.

Curative treatment is required to treat cancers, heart disease and stroke. This is much more expensive than preventative health care and may involve lengthy hospitalization.

There are many benefits of preventing diseases—treatment, lost productivity and health-care costs are major burdens to the economy, businesses and families.

Many of the world's poor are at increased risk of disease. Many lack information, money or access to health facilities for adequate health care. The poor may be socially marginalized, and may have to make difficult choices.

Test yourself

F.15 Describe the advantages and disadvantages of in vitro food.   [2+2]

F.16 Outline the main advantages of genetically modified food.   [3]

▲ **Figure F.4.1.** Regular exercise is an effective preventative treatment

The provision of health-care services may be public, for example, the UK's National Health Service, or it may be private, as is largely the case in the USA. However, not everyone can afford private health-care.

Managing pandemics

Pandemics are global epidemics. Their large scale makes them difficult to manage and they may involve new diseases or relatively unknown ones.

Following the outbreak of Ebola in West Africa in 2015, the Nigerian government established a massive public health campaign. Containment was the key to ending Ebola. Everyone who had been exposed to the virus was found and monitored, and isolated if they developed the symptoms. Television broadcasts and social media were used to reassure people. Gatherings were banned. Markets and schools were closed, and school lessons were given over the radio. Although Ebola had the potential to become a pandemic, it did not happen due to the speedy response of health officials not only in West Africa but in other areas such as North America and Europe.

Case study: The diabetes pandemic

The number of people worldwide with diabetes is around 422 million, a figure likely to double by 2035. Diabetes is a chronic, lifelong condition and a major cause of blindness, kidney failure, heart attacks, stroke and lower limb amputation. The disease reduces both a person's quality of life and life expectancy.

Four major trials have demonstrated that lifestyle changes involving diet (reducing fat intake), preventing obesity and increasing physical activity can delay or prevent type II diabetes among people at high risk.

Test yourself

F.17 Explain why a pandemic could trigger a global recession. [3]

F.18 Define the factors that led to the successful containment of the Ebola virus. [2]

QUESTION PRACTICE

The following diagram shows factors affecting food insecurity.

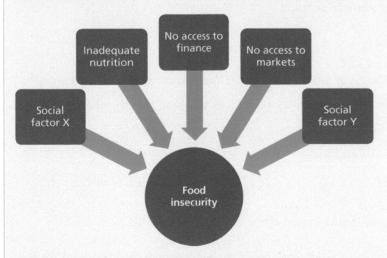

a) **Identify** what social factors X and Y could be. [2]

b) i) **Outline** why two named groups of people experience food insecurity. [2 + 2]

   ii) **Explain** two ways in which improved food security can be achieved. [2 + 2]

Essays

Either: Examine the impact of one vector-borne disease. [10]

Or: Examine the nutrition transition that occurs as countries develop. [10]

How do I approach these questions?

a) 1 mark is awarded for each valid suggestion, with a maximum of 2 marks.

b) i) You need to outline two different population groups that are at risk of food insecurity, and give some detail for each one. Describe the conditions that mean they have limited food supply—it is not good enough to say that they have "no food"; you should think of different reasons why their food supply is limited.

ii) This asks for ways in which food security can be achieved. Each of the reasons should be developed or demonstrated using an example.

Essays

For these essays, the command term is examine. This means that you should discuss the underlying assumptions and interrelationships of the issues presented in the question. For the first essay, you should consider a number of impacts of a named vector-borne disease on people and societies, and relate these to the four key concepts (places, processes, power and possibilities). For the second essay, you should examine how people's diet changes as they transition from LIC to MIC to HIC. Remember to refer to specific examples, and some critical thinking and evaluation should be present in your answer.

SAMPLE STUDENT ANSWER

a) Other social factors affecting food security include having to walk long distances to get water and firewood and poor education.

Marks 2/2

b) i) One population group experiencing food insecurity is people living in Syria. This is because there is a civil war going on there and food supplies are disrupted. Another population group experiencing food insecurity are people in Sudan. This is because there is a famine there, and farmers are unable to produce many crops.

▲ 1 mark

▲ 1 mark

▲ 1 mark—very simple, but acceptable

Marks 4/4

▲ 1 mark

▲ 1 mark

ii) One way of producing more food is through the use of GM crops. In vitro farming also helps produce much food from a single stem cell.

Neither point developed.

Marks 2/4

Essays

Either: Examine the impact of one vector-borne disease.

▲ Identifies a valid vector-borne disease

Malaria is a life-threatening disease of humans caused by the plasmodium parasite and transmitted to people via the bite of the female Anopheles mosquito. In 2015, around 100 countries and territories had ongoing malaria transmission. About 3.2 billion people – almost half the world's population – are at risk of malaria. However, between 2000 and 2015, malaria incidence among populations at risk (the rate of new cases) fell by 37 per cent globally. At the same time, malaria death rates in populations at risk fell by 60 per cent globally among all age groups and by 65 per cent among children under five. Sub-Saharan Africa carries a disproportionately high share of the global malaria burden. In 2015 the region suffered 88 per cent of malaria cases and 90 per cent of malaria deaths. The direct cost of malaria to individual households includes medication, doctors' fees and preventative measures such as bed nets, which help to reduce transmission. Infected individuals are unable to work, which can reduce family incomes during the attacks.

Some population groups are at considerably higher risk of contracting malaria, and developing severe disease, than others. These include infants, children under five years of age, pregnant women and patients with HIV/AIDS, as well as non-immune migrants, mobile populations and travellers. Children with severe malaria frequently develop one or more of the following symptoms: severe anaemia, respiratory distress in relation to metabolic acidosis, or cerebral malaria.

▲ Good detail

▲ Evaluative comment

▲ Shows a trend

▲ Identifies the worst-affected area and gives some support

▲ Good opening paragraph —sets the scene well and identifies changing incidence/death rate

▲ Death has been covered in the first para. Economic costs now covered

▲ Demographic focus

▲ Social costs—identifies some of the "at risk" population

▼ Not backed up with real-life examples

The first paragraph was full of detail but thereafter it becomes quite generic. A named, located example (such as Nigeria or DR Congo) would be good to show the changes/impacts in a real-life situation.

Marks 7/10

Or: Examine the nutrition transition that occurs as countries develop.

As income increases in low-income countries (LICs), there is an increase and a change in food consumption patterns. People in LICs generally derive their food energy mainly from carbohydrates, while the contribution of fats is small and that of meat and dairy negligible. In Bangladesh, for example, an LIC, people derive 80 per cent of their nutritional energy from carbohydrates and 11 per cent from fats. People in high-income countries (HICs) generally derive most of their food energy

▲ Sound definition

▲ Good description

▲ Real-life example

▲ Very good introduction—clear and sets the scene

▲ True for modern civilizations—some would say there were earlier NTs

▼ Should state that this is for LIC and MICs. HICs have not been increasing substantially since the 1970s.

▼ Good range of reasons, but it would be good to develop some of these with examples

▲ Good point, a kind of reverse-transition in HICs

▲ Sound comparison—it would be interesting to have details for particular societies

▲ Good conclusion—brings the main points together and has a brief evaluative comment

from carbohydrates and fat, with a substantial contribution from meat and dairy. The average consumer in the US, France and Denmark, for instance, derives 45–50 per cent of their food energy from carbohydrates and 40 per cent from fats.

Studies of human nutrition have shown that worldwide a nutrition transition is taking place, in which people are shifting towards more affluent food consumption patterns. The nutrition transition began in developed countries 300 years ago. It coincided with great economic growth. For LICs, a small increase in income may lead to a large increase in calorie intake, while for HICs increases in income may not lead to an increase in calorie intake. Food consumption per capita has increased substantially since the 1970s (both in energy and protein content). Growth rates are consistently higher in LICs, but their consumption levels per capita are still much lower than in HICs. The transition in diet is mainly influenced by higher income per capita – but food prices, individual and sociocultural preferences, refrigeration, and other concerns also play a role.

In HICs, the main dietary changes since the 1970s have been the reduction in cereals, while mainly vegetable oil and, to a smaller extent, meat intake increased. Animal protein intake has been stabilizing: an increasing part of the population seems to be interested in reducing/replacing it for various reasons (ethical, health-related, environmental and economic).

In LICs the diet has diversified since the 1970s. Intake of cereals, including rice, as well as vegetable oil, sugar, meat and dairy is higher now compared to 1970s, although in more recent periods cereal intake has stagnated and even declined. Their share of cereals also exceeds the share in HICs.

There is a strong positive relationship between level of income and consumption of animal protein and a negative relationship with staple foods. In LICs, dairy, fish and pulses are driving increases in total protein availability per capita. Sugar intake is also stabilizing. These numbers seem to suggest that the diet in LICs is slowly evolving in the direction of the HICs, with the exception of sugar.

Good overall account—apart from the examples in the opening paragraph, most of the support relates to HICs/LICs.

Marks 8/10

G URBAN ENVIRONMENTS

According to the UN, much of the world's population lives in urban environments. These areas are constantly evolving as people enter and leave. This creates opportunities and challenges economically, socially and environmentally for local and national governments, infrastructure and for the people who live in these places.

You should be able to show:

✔ the characteristics and distribution of urban **places**, populations and economic activities;

✔ how economic and demographic **processes** bring change over time to urban systems;

✔ the varying **power** of different stakeholders in relation to the experience of, and management of, urban stresses;

✔ future **possibilities** for the sustainable management of urban systems.

G.1 THE VARIETY OF URBAN ENVIRONMENTS

You should be able to show the characteristics and distribution of urban places, populations and economic activities:

✔ Characteristics of urban places, including site, function, land use, hierarchy of settlement (including megacities) and growth process (planned or spontaneous);

✔ Factors affecting the pattern of urban economic activities (retail, commercial, industrial), including physical factors, land values, proximity to a central business district (CBD) and planning;

✔ Factors affecting the pattern of residential areas within urban areas, including physical factors, land values, ethnicity and planning;

✔ The incidence of poverty, deprivation and informal activity (housing and industry) in urban areas at varying stages of development.

- **Site** – the actual land on which a settlement or an urban area was established.

- **Function** – the main economic activities that take place in an urban area.

- **Deprivation** – people and groups that experience a lower standard of living than the majority of people living in an urban environment.

- **Informal housing** – residential areas that have been built illegally by residents.

- **Informal sector** – people who work in the informal sector do not declare their income and pay no tax on it. This is also known as the black economy, the shadow economy or the grey economy.

Characteristics of urban places

When considering the original sites of settlements, the presence of flat land would have allowed for the straightforward construction of buildings, and the proximity to water would provide a supply for drinking and for irrigating crops.

Concept link 🔗

PLACES: Urban environments evolve, and towns and cities develop an identity as a whole along with the areas within them. Economic and social processes take place and the physical geography of a locality will also enable change.

The location of a settlement would subsequently give a village, town or city a function. For example, a market town could develop where agricultural produce from nearby rural areas could be bought and sold.

A settlement can be multifunctional and it can evolve over time. The land use in a town or city can also vary, with industrial, residential and recreational land uses existing in many urban environments. Planning regulations may dictate how land is to be used in an urban area and also how it is not be used.

Urban growth can be spontaneous (for example, illegal settlements) or it can be planned (such as new private or public housing developments). Thus urban places are unique and dynamic as they are established and evolve.

Factors affecting the pattern of urban economic activities (retail, commercial, industrial)

Secondary (manufacturing) and tertiary (services) economic activities are located in urban areas. Secondary activities will require a larger amount of land compared to tertiary economic activities. All land has a value, and in theory the closer you get to the central business district (CBD), the higher the cost of the land. Service industries such as offices, restaurants and bars tend to be located closer to the CBD as workers and consumers will all have access to this area via public and private transportation.

Factors affecting the pattern of residential areas within urban areas

Residential areas have already been mentioned above in relation to the establishment of informal housing. The location of formal housing is decided based on the cost of the land and the planning restrictions set by the local authorities. Land that is more expensive and closer to the CBD will be used for apartment blocks, since building upwards rather than outwards (which might not be possible anyway) will reduce the development costs in terms of the purchase price of the land.

Redevelopment can take place. For example, apartment blocks may be demolished and replaced with detached/single-family houses.

Housing can be both privately owned and public (provided by the government), and the location of public housing can vary. For example, inner-city public housing and edge-of-city public housing can exist in cities in the same country. The value of the housing can also vary, with some parts of an urban area being more desirable than others.

As people migrate to urban environments, the diversity of places increases. Some areas may have a higher concentration of people from a particular ethnic background due to the existence of familiar cultural traits or cheap rent prices.

In summary, a range of processes creates patterns of different housing types in an urban area.

G.2 CHANGING URBAN SYSTEMS

You should be able to show how economic and demographic processes bring change over time to urban systems:

✔ Urbanization, natural increase and centripetal population movements, including rural–urban migration in industrializing cities, and inner-city gentrification in post-industrial cities;

✔ Centrifugal population movements, including suburbanization and counter-urbanization;

✔ Urban system growth including infrastructure improvements over time, such as transport, sanitation, water, waste disposal and telecommunications;

   ✔ Case study of infrastructure growth over time in one city;

✔ The causes of urban deindustrialization and its economic, social and demographic consequences.

Urbanization, natural increase and centripetal population movements

Urbanization has taken place over time; both middle-income countries (MICs) and low-income countries (LICs) have experienced significant urban growth in population over recent decades. Rural-to-urban migration, a centripetal movement, has increased the amount of people living in urban areas. Migrants may then have children once they are settled in the urban area, thus the processes of migration and natural increase (the difference between crude birth and death rates) can increase the population.

Gentrification is a process that has been increasingly highlighted in many towns and cities, as wealthier people move into an area creating significant and sometimes controversial economic and social effects.

Centrifugal population movements

Centrifugal population movements are the opposite of centripetal movements as people move away from the centre of urban areas via processes such as suburbanization and counter-urbanization.

Urban system growth

Urban areas should be acknowledged as systems, and as such they have inputs, processes and outputs. The inputs, for example, could be the in-migration of people. The processes might be the movement of people via public and private transportation. Outputs can be waste such as rubbish, and the management of the outputs can determine the level of sustainability within an urban system. The ability for an urban environment to cope with change within the system, such as an increase in people or a need to reduce the level of pollution, can create issues and challenges for different stakeholders.

- **Centripetal** – movement towards an urban area.

- **Centrifugal** – movement away from an urban area.

- **Gentrification** – a general term for the arrival of wealthier people in an existing urban district, a related increase in rents and property values, and changes in the district's character and culture.

- **Post-industrial city** – a city whose economy has shifted from producing goods and products to one that mainly offers services.

- **Deindustrialization** – the process of social and economic change which is due to the reduction in industrial capacity or the activities of a country's manufacturing and heavy industry.

- **Urbanization** – an increase in the proportion of people living in towns and cities compared to rural areas.

- **Counter-urbanization** – a movement of people away from urban areas to rural areas and smaller settlements.

Content link

Connect this information with the population changes described in unit 1.

Test yourself

G.5 Identify one type of centripetal movement. [1]

G.6 Discuss the process of gentrification. [6]

G.7 Explain why cities in some parts of the world have higher rates of population growth than others. [2+2]

Test yourself

G.8 Analyse the environmental consequences from city growth relating to solid waste in São Paulo.   [4]

Case study: Infrastructure growth over time in São Paulo, Brazil

São Paulo is a city within the state of São Paulo and it is one of the largest cities in the world. In 2018 the population of the metropolitan urban area was estimated to be 21,730,000. While the total fertility rate (TFR) is below replacement level (1.69), the city has grown due to previously higher fertility rates and rural-to-urban migration which began in the mid-19th century.

The city has had issues with transport, sanitation and water. Congestion is a major issue which increases the urban stress for those travelling in and around the city. With one car for every two people, the city's road networks have not grown at the same pace as car ownership.

Water is sourced from outside the city. This system has been described as inefficient due to leaks which mean that additional water has to be sourced from elsewhere to meet the needs of residents. In addition, recent drought led to 12-hour water cut-offs for many of the city's population and reservoirs fell to very low levels. Both the state of São Paulo and the city have struggled to treat sewage, and waste has entered rivers, reservoirs and coastal waters. Two of São Paulo's main rivers, the Tietê and the Pinheiros, are in the top 10 most polluted rivers in Brazil.

Each citizen produces approximately 1.1 kg of waste per day and most of this waste is deposited in landfills. Teams of garbage collectors (*catadores*) travel around the city to collect waste that can be recycled.

The causes of urban deindustrialization and its economic, social and demographic consequences

As already discussed, cities are dynamic with a constant flux of people arriving and leaving, and industrial areas of a city are also subject to change. Cities in high-income countries (HICs) have seen the demise of heavy industry in cities and the relocation of secondary industry to MICs. The loss of industry, or deindustrialization, in a city can have economic, social and demographic consequences.

Economic consequences may include the loss of employment which can then create a cycle of poverty if former employees find that there are no alternative jobs in their urban area or they do not have the required skillset or retraining opportunities. The loss of income can create health problems, such as depression, as people feel that they can no longer support themselves and their families. Crime and social unrest may increase, causing people to leave an area in search of new employment. So not only will the area have lost industry and jobs, there will also be a reduction in the number of people living there. Alternatively, renovation could take place, and old mills and factories could be converted into apartments.

The process of deindustrialization creates a range of consequences as places evolve.

G.3 URBAN ENVIRONMENTAL AND SOCIAL STRESSES

You should be able to show the varying power of different stakeholders in relation to the experience of, and management of, urban stresses:

✔ Urban microclimate modification and management, including the urban heat island effect, and air pollution patterns and its management;

　✔ Case study of air pollution in one city and its varying impact on people;

✔ Traffic congestion patterns, trends and impacts;

　✔ Case study of one affected city and the management response;

✔ Contested land-use changes, including slum clearances, urban redevelopment and the depletion of green space;

　✔ Detailed contrasting examples of two affected neighbourhoods and their populations;

✔ Managing the impacts of urban social deprivation, including the cycle of deprivation and geographic patterns of crime;

Urban microclimate modification and management

An urban microclimate is an urban area that has a climate that is different to the surrounding rural area. Towns and cities are often warmer than their surrounding areas due to the urban heat island effect, as tall buildings and dark surfaces retain heat from solar radiation. There will also be more rainfall as there is a greater amount of dust particles upon which water vapour can condense. Wind speeds vary more due to the layout of buildings in relation to prevailing winds. The large number of vehicles and higher frequencies of congestion in urban areas will also increase the amount of pollution compared to rural areas, especially when there is less vegetation to filter the air.

Test yourself

▼ Figure G.3.1. Urban heat island profile

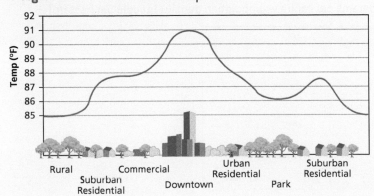

G.9 **Define** the term albedo. [2]

G.10 Use figure G.3.1 to **describe** how the temperature changes between the rural area and the suburban residential area. [3]

- **Albedo** – the amount of incoming solar energy reflected back into the atmosphere by the Earth's surface.

- **Microclimate** – the distinctive climate of a small-scale area, such as a garden, park, valley or part of a city.

- **Urban heat island** – an urban area where the temperatures are higher than the rural areas surrounding it.

- **Slum clearance** – the demolition of slums, sometimes accompanied by the rehousing of the inhabitants, to improve living conditions and the environment of an inner city.

- **Cycle of deprivation** – The persistence of poverty and other forms of socio-economic disadvantage through generations via a sequence of events.

Concept link 🔗

POWER: Every person living in an urban environment contributes to the social and environmental well-being of that place. For example, the collective will of residents, industry, and city authorities have the power to try and control stress to maintain and improve the quality of life. As cities change, it is necessary to re-evaluate the choices that these different stakeholders make. For example, further reducing the amount and type of vehicles on the place's road network.

>> **Assessment tip**

Ensure that data or quantification is included when describing charts or diagrams. The inclusion of data will not necessarily earn you a mark, but it will often be necessary in order to gain the total marks available for a question.

G.11 Choosing either rural, downtown or urban residential, **justify** how human activities can either increase or decrease the effects of an urban heat island. [2+2]

G.12 Apart from temperature, **justify** how human activities can modify the microclimate of an area. [3+3]

Air pollution is much higher in urban areas than in rural areas, but cities have differing levels of air pollution, for example, Mexico City compared to Vancouver. Various pollutants can be present from vehicle emissions and from industry. $PM_{2.5}$ and PM_{10} (particle matter 2.5 and 10 micrometres respectively) can get into a person's lungs and pass into the bloodstream, causing breathing problems and lung cancer. The World Health Organization states that 20 micrograms per cubic metre of air is an annual average, but in some cities the average PM_{10} is over 300 micrograms per cubic metre.

Case study: Air pollution in Onitsha, Nigeria

The city of Onitsha in Nigeria has an annual pollution reading of 594 PM_{10}, which is one of the highest in the world. This is a result of emissions from vehicles as well as from industry (mining, manufacturing cement). Dust storms that occur in the region also generate finer particles such as $PM_{2.5}$. There is currently limited evidence about the impact on people's health, but it is anticipated that air pollution will be a major cause of premature death in Onitsha in the coming years.

Traffic congestion patterns, trends and impacts

Case study: Traffic congestion in Mexico City

Mexico City has been plagued by traffic congestion for decades and this has contributed to the environmental and social stress that exists there. The physical geography and prevailing winds ensure that the emissions from vehicles plus those from industry to the north of the city remain in the "bowl" in which the city is located, surrounded by volcanoes.

Several management strategies have been implemented, such as the introduction of the Metrobus-dedicated bus lane, to encourage people to avoid driving into the city by providing efficient public transportation.

Test yourself

▼ **Figure G.3.2.** Transport emissions in kilograms per capita vs population density (people per hectare)

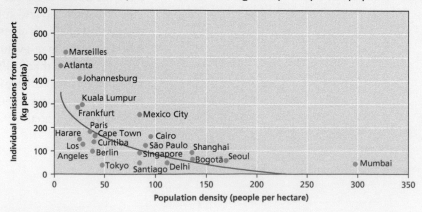

Source of data: World Bank (2009)

G.13 Describe the relationship between population density and air pollution from transport. [2]

G.14 Compare and contrast the air quality in two cities of your choice. [6]

Contested land-use changes

Slum clearance takes place in both HICs and LICs, and the term "slum" can relate to illegal settlements in LICs and dilapidated 19th-century housing in cities in HICs. Slums in LICs are often demolished by the city authorities since they have been illegally built or the land has been sold to be developed. Slums in HICs have been redeveloped into new public and private housing developments. Redevelopment can often be controversial, for example, if the new housing does not provide affordable homes.

Case study: Depletion of green space in Sydney

In Sydney, almost 75% of all new residential developments between 2011 and 2030 will be constructed at "in-fill" sites within the city's boundaries. Cook Cove is one such site and local campaign groups have been protesting about the loss of playing fields, cycle tracks, community playgrounds and the wetlands in order to build 10,000 homes. There are also plans to relocate a golf course to the park, which is green space of course, but this has been met with protests since the multi-use public park will be replaced by a private access only, single-use golf course.

Case study: Redevelopment in Mumbai

In Mumbai, the redevelopment of the Dharavi slum has attracted controversy. As part of the redevelopment, those living in the slum would be entitled to a home measuring 350 square feet, which is smaller than some of the current homes constructed by Dharavi's residents. The new development will involve building upwards, whereas people currently live in homes that are on the ground level. In addition, there is no plan for the provision of space for the informal industry that is currently present. Therefore many local people are not happy with the plans.

▼ Figure G.3.3. Huts in the Dharavi slum

Managing the impacts of urban social deprivation

Urban environments are areas with significant inequality. Economic and social differences exist, with some residents marginalized in cities at varying levels of development.

In many countries, local and national governments have the responsibility of trying to improve the quality of life in these areas and to break the cycle of poverty. In the borough of Newham in London, for example, which is one of the most deprived parts of the city, the council created a team that identified people living in low-quality housing, such as in garden sheds that had been converted. A strategy called Workplace was also developed in which people were able to attend training courses paid for by the local authority which resulted in those people finding work as a result of their new skills.

The power of developers and politicians can be very influential in changing land use when developing urban areas.

Test yourself

G.15 Choosing a particular stakeholder, **discuss** how they have the power/responsibility to resolve a social or environmental stress in urban areas. [1+4]

G.4 BUILDING SUSTAINABLE URBAN SYSTEMS FOR THE FUTURE

- **Resilient city design** – a city that has been designed to absorb future shocks and stresses to its social, economic and technical systems and infrastructures so that it can maintain essentially the same functions, structures, systems and identity.

- **Geopolitical risk** – the risk from a government or an organization in one country influencing an urban area's policies in another country.

- **Urban ecological footprint** – the theoretical measurement of the amount of land and water that an urban population requires to produce the resources it consumes and to absorb its waste under prevailing technology.

- **Smart city design** – the effective integration of physical, digital and human systems in the built environment to deliver a sustainable, prosperous and inclusive future for its citizens.

- **Retrofitting** – the directed alteration of the fabric, form or systems that comprise urban environments to improve energy, water and waste efficiencies.

Content link

This section of the syllabus connects with unit 1, since it examines changing amounts of people living in urban areas. The inclusion of content from units 1–3 (paper 2) and units 4–6 (paper 3) is perfectly valid in order to develop your exam responses for paper 1 options.

You should be able to show examples of future possibilities for the sustainable management of urban systems:

✔ Urban growth projections for 2050, including regional/continental patterns and trends of rural–urban migration, as well as changing urban population sizes and structures;

✔ Resilient city design, including strategies to manage escalating climatic and geopolitical risks to urban areas;

   ✔ Two detailed examples to illustrate possible strategies;

✔ Eco-city design, including strategies to manage the urban ecological footprint;

   ✔ Two detailed examples to illustrate possible environmental strategies;

✔ Smart city design and the use of new technology to run city services and systems, including purpose-built settlements and retrofitting technology to older settlements.

Urban growth projections for 2050

Each year, the United Nations produces its World Population Prospects report which details current and future demographic trends based on past and current data. A greater number of the world's population is now living in urban areas. In 2018 the UN reported that 70% of the world's projected population in 2050 (10 billion) will be living in urban areas.

Test yourself

▼ **Figure G.4.1.** Past and future urban and rural populations: North America, Europe and Oceania (data from 2014)

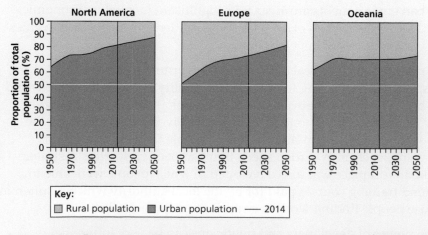

Source of data: Department of Economic and Social Affairs, United Nations (2014)

G.16 Define the term urbanization. [1]

G.17 Using figure G.4.1, **identify** the continent with the highest proportion of people living in urban areas in 2014. [1]

G.18 Identify the continent with the projected lowest rate of urbanization between 1950 and 2050. [1]

G.19 Describe the projected change in the proportion of people living in urban areas in Asia between 1950 and 2050. [3]

> **Concept link**
>
> **POSSIBILITIES:** Towns and cities are being retrofitted in order to respond to climatic and geopolitical risks, while new urban environments are being planned and designed to ensure that environmental sustainability is achieved.

Resilient city design

The UN reported in 2016 that 90% of the world's cities were located in areas that are vulnerable to natural disasters relating to climate such as flooding, drought and cyclones. Cities need to have plans in place to withstand these climate-related risks.

Possible environmental strategies: Rotterdam and Singapore

Rotterdam, in the Netherlands, is vulnerable to flooding from three rivers and from sea-level rise from the North Sea. Managed flooding from these rivers is allowed in some public areas which contain sunken voids in the ground made of concrete that can be filled with water. The Maeslant Barrier is positioned on the River Rhine and it protects the city from a storm surge from the sea.

In Singapore, a city state that is vulnerable to tropical storms, hard walls have been built to protect against storm surges, and if any land is reclaimed for construction it must be at least 4 metres above sea level compared to a previous level of 3 metres.

Geopolitical risks to urban areas can also involve conflict. Instability that exists in nearby places can spread to cities, especially when there are historical or cultural connections between places. For example, since 2011, refugees have spread from Syria to neighbouring countries, such as Jordan, as well as European cities, such as Cologne, which puts pressure on national and city governments to accommodate this diaspora.

Eco-city design

In general, there has been an increase in environmental awareness and cities are striving to implement city designs in order to become more environmentally sustainable. This is beneficial to the urban ecological footprint as outputs are recycled.

Smart city design incorporates technology in order to run city services and systems in a more efficient manner. Many different aspects of a city can be affected when new technology is introduced, such as cyber defence, mobility, renewable energy, etc.

There is a range of possibilities for improving urban areas.

Possible environmental strategies: Milan and Oregon

Greening of a city is a strategy that has been implemented by Milan, Italy. One of the famous examples, Bosco Verticale ("vertical forest"), consists of two apartment blocks that contain 800 trees and a range of other vegetation equating to 20,000 square metres of forest which absorbs carbon dioxide and dust particles.

> **Content link**
>
> Environmental sustainability in societies is discussed further in unit 3.3.

▼ **Figure G.4.2.** A shaded street in Masdar City, a planned eco-city project in the UAE

Portland, Oregon, has been a pioneer amongst US cities in developing green initiatives by creating 188 miles of cycle paths, and almost 10% of the population commutes via bicycle. It has also invested significantly in public transportation with an extensive bus and tram network, and the city aims to be powered completely by renewable energy by 2050.

>> **Assessment tip**

Since the ecological footprint is calculated using several different variables, there are a number of different answers to question G.21. A good approach would be to discuss land and water for one part of the answer and then to discuss the way in which waste is managed for the second part of the answer.

>> **Assessment tip**

If question G.21 was in the middle of the structured questions, then it would be a good opportunity to include an example that you have studied. Marks are available for developing descriptions and explanations via extension, and for the inclusion of examples. The Brazilian city of Curitiba would be an ideal example of a city that implemented various measures to reduce its urban ecological footprint.

Test yourself

G.20 Define the term urban ecological footprint.   [2]

G.21 Analyse how the urban ecological footprint can increase or decrease.   [3+3]

QUESTION PRACTICE

Examine figure 6.3.3, on page 103, which shows the Dharavi slum in the city of Mumbai.

a) **Identify** two pieces of evidence showing that this is an informal residential area, other than the poor quality of housing.   [1 + 1]

b) Using an example of a city you have studied, **outline** two reasons why certain ethnic groups tend to be located in specific places.   [2 + 2]

c) **Explain** two processes that are responsible for population growth in megacities.   [2 + 2]

Essays

Either: Using examples, **examine** the varied effects of human activity on urban microclimates.   [10]

Or: Examine the similarities and differences in patterns of urban deprivation for two or more cities you have studied.   [10]

How do I approach these questions?

a) The question states that you are not permitted to discuss the construction of the dwellings. Therefore you should write about other aspects of how the place has developed in your answer.

b) You should have studied an example that demonstrates why different groups tend to locate in certain areas of an urban settlement. Consider a range of social reasons in your answer.

c) Further explanation is needed for this question and you must recall the fundamental reasons why population in any place increases or decreases.

First essay choice:

This essay requires an understanding of the different climatic variables. So there is plenty of opportunity to include a wide range of appropriate terminology which will increase your mark for knowledge and understanding. Your introduction should define the term "urban microclimate", and you should be aware that human activity does not always have a negative impact on urban microclimates.

Second essay choice:

The question requires information about the distribution of deprivation in different cities, so this must be discussed in the answer. Evaluation is expected, so consider how the location of deprivation is either similar or different in cities in countries at different levels of development. The introduction should define important terms in the question, such as deprivation and development, while the main body of the essay should explain the causes behind the development of these areas of deprivation. You should contextualize the level of deprivation in relation to the socio-economic status of each city and consider what the characteristics of these places are that classify them as deprived.

SAMPLE STUDENT ANSWER

a) There is a lack of infrastructure present such as official roads and pavements. Also, the development has not be planned since it is haphazard in nature and does not follow a urban plan, for example.

▲ Two pieces of evidence from the figure given

Marks 2/2

b) Most cities will have areas where certain ethnic groups tend to reside in an area. Amsterdam Zuidoost is area of Amsterdam that contains a high concentration of people from Suriname because the cost of rent was very cheap when people moved from Suriname. In addition, when people move from Suriname to Amsterdam, they may prefer to live in an area that contains other people from Suriname because there is more chance that the same cultural traits may be seen, such as being able to purchase Surinamese food such as Moksi Meti.

▲ 1 mark—valid city and area within the city

▲ 1 mark for a valid reason

▲ Another valid reason given

Two distinct and developed reasons have been provided by using a place-specific example.

Marks 4/4

c) Population increases or decreases due to changes to a city's fertility rate. More children will mean more people and vice versa. The second reason is due to the death rate. If more people are dying then this change the population.

▼ 2 marks—the natural increase or decrease is discussed in the answer. Population changes due to natural and migration and the latter should also be discussed with reference to in-migration and out-migration

Marks 2/4

Essays

Either: Using examples, **examine** the varied effects of human activity on urban microclimates.

▲ Appropriate knowledge

▲ Appropriate terminology

▲ The introduction shows knowledge and it provides a foundation for discussing temperature, air quality and wind

Human activities in urban environments cause the formation of an urban microclimate. The different human activities in urban environments have different effects on the climate. Heavy polluting industries and transport lead to the creation of smog. The construction of buildings and the release of particulate matter lead to the urban heat island effect. The construction of buildings also disrupts wind patterns, causing the Venturi effect. These different effects combine to create the unique microclimate experienced by urban areas.

▲ Shows knowledge

The urban heat island effect is the formation of an area of higher temperature in the urban island which is surrounded by rural areas that have a lower temperature. This effect is caused by a variety of different human activities. Many human activities such as transport and industry result in the release of particulate matter into the atmosphere. This particulate matter

▲ Appropriate terminology

can then act as cloud condensation nuclei as it allows water to condense around it, which will increase the cloud coverage raising the possibility of increased rainfall. In addition, the cloud coverage will trap solar insolation raising the temperature. Humans also raise urban temperatures due to the buildings and infrastructure in urban areas. In many inner city areas, concrete and tarmac are used in the construction of buildings

▲ Appropriate terminology

and roads. This lowers the albedo of the area as concrete absorbs 60% of solar insolation, meaning that more solar insolation is absorbed so it can be released at night, increasing the temperature. Recently this has changed slightly as in more modern cities the central business district contains many

▲ Explanation present

buildings made mainly of glass that increases the albedo. This means that more solar insolation is reflected during the day, but this will then be trapped by the high cloud coverage causing

▲ A focused paragraph that discusses the causes of temperature increase

temperatures to increase during the day. Many different human activities result in the release of heat into the atmosphere as heat escapes from residential buildings.

The effect of human activities on the urban microclimate is so great that in cities such as London the average temperature is two degrees above that of the surrounding rural area.

Human activities affect the urban microclimate through the creation of photochemical smog due to the emissions from cars and other vehicles. The smog affects the climate as it can act as cloud condensation nuclei increasing cloud coverage. This photochemical smog has had a big effect on the temperature as it can react with other chemicals in the atmosphere. Nitrous oxides can cause ozone depletion as they catalyse the photodissociation of ozone, meaning that more solar insolation can reach the urban area, increasing the temperature. Human activity also affects winds patterns in urban environments, leading to greater wind speeds and more fluctuating wind directions due to the Venturi effect. In cities, the normal route for wind is blocked by different buildings, so wind is channelled through gaps in buildings increasing the wind speed. However, wind is also refracted around the sides of buildings meaning that the wind is more disordered and chaotic in cities. This effect causes wind speed to be three times higher in urban areas, as seen in Chicago, nicknamed "the windy city".

Human activities of transport, buildings and industry have an important effect on the urban microclimate. The different activities affect different areas of the climate and to a different extent, but they eventually result in urban areas having higher temperatures, wind speed and precipitation. The effect of these activities varies in different cities and while they may be welcomed in some for providing a more comfortable climate, they can cause significant effects in others. The impact of this microclimate can exacerbate issues such as with the Paris heatwave of 2003 where 2,000 people died from heat-related issues and Tokyo in 2018 where over 50 people died. The impact of the urban environment on the climate is changing, however, as industries expand but alter slightly, such as the high albedo of the CBD or the release of photochemical smog.

▼ Only one example has been included with limited detail

▼ This needs clarification, and it repeats what has been said above

▲ Knowledge

▲ More relevant knowledge and understanding although no examples provided

▼ There should be a new a paragraph here since the next section discusses wind speed and direction

▲ A second example is included although again, limited in detail. The response includes some relevant knowledge and understanding

▲ The conclusion summarizes the content of the response. However, some of these relevant examples could have been included earlier in the response

This is a very good response which addresses a range of different factors that can be influenced by human activity. There is description and explanation with brief examples present. The structure could be improved slightly by having a greater focus for each paragraph. Evaluation could be been included by discussing how human activities have reduced the amount of pollution or temperature in cities. This would have increased the mark.

Marks 8/10

Or: Examine the similarities and differences in patterns of urban deprivation for two or more cities you have studied.

Deprivation can be defined as people experiencing a low quality of life due to a lack of or low level of income and health issues or a low life expectancy. Urban areas, being unequal, will have some parts that contain people with high levels of deprivation and other parts with low levels. Development refers to the economic development of a city or country or the level of their HDI. The statement is somewhat correct in that cities at different development levels will have areas that are deprived, but the location of these will differ. This will be demonstrated by discussing Lagos, London and Paris.

In LICs and MICs, areas that are deprived tend to be located towards the outskirts of cities. Illegal settlements are established, quite often by rural to urban migrants who construct their own homes. In Lagos, the shanty town of Makoko where 250,000 people live, is located where the land meets the water on the edge of the city. The land is swampy and many of the homes are built on stilts to avoid the water. The children that live there tend to drop out of school at a young age in order to try and bring in money via fishing and there is a lack of government investment. In fact, they are under constant threat of their homes being removed by the government. No basic services are provided such as clean water and electricity and most residents live below the extreme poverty line of $3 a day. This example demonstrates that deprivation is located on the outskirts of a city, on land (and water) that is undesirable.

In London, one of the poorest districts is called Tower Hamlets and unlike Makoko, it exists in the inner-city rather than the outer-city. It is the most deprived part of London with just less than 50% of children living in poverty and an unemployment rate of 8%. Here though there is government investment which attempts to improve the socio-economic status of the residents of Tower Hamlets.

▲ The introduction contains a very good level of knowledge with important terminology defined, a clear thesis statement and evidence of structure with the examples listed to be discussed

▼ Quite a sweeping statement when including both LICs and MICs

▲ Appropriate example

▲ There is plenty of evidence in this paragraph for the location of deprivation and the characteristics of the deprivation

▼ It is actually US$1.90 a day for the extreme poverty threshold

▲ Connects with the thesis statement, a well-structured paragraph

▼ Refers to location although could be more explicit in order to fully connect with the essay statement

▲ Detail is provided for this example

It should be noted that the definition of poverty used in Lagos is different to London. In Lagos it is absolute poverty, living on less than $3 a day whereas in London, the level of poverty is 'relative poverty' which means that peoples' income is 60% below the UK's median income. This means that when comparing areas of poverty in both cities, the type of poverty is different. Therefore not only is the location of poverty in both cities different, but the level of deprivation in both cities is different also. A similarity though is that both cities have levels of inequality and that there are clear areas where people are more deprived than other parts of the city. Finally there is a similarity between the pattern of deprivation in cities in MICs and LICs in terms of deprivation being on the outskirts. Paris has a number of 'sink estates' which are located on the outskirts of the city. The centre of Paris contains some of the most expensive real estate in the world whilst the sink estates contain government-provided housing or private rented accommodation that is in poor condition. People on these estates feel marginalized and forgotten about by the government. In summary, patterns of deprivation are different and similar in cities with different levels of development.

▼ This figure is inaccurate

▲ Knowledge is demonstrated

▲ This paragraph includes explanation which extends the explanation of the difference between two cities

▲ Clear point made linked to the thesis statement

▲ Appropriate example

▲ Some evidence provided for deprivation

▼ The evidence of deprivation could be more detailed—this information could relate to any city

▼ The conclusion is very brief and the points that have been made in each paragraph should be included

Apart from the inaccurate reference to the extreme poverty level, this essay contains a high level of knowledge and each paragraph contains explanations linked to the thesis statement and question. Evaluation is present since similarities and differences are justified via the well-chosen examples. Some paragraphs contain more explanation than other paragraphs whilst the conclusion is very brief, which prevents the response from receiving full marks.

Marks 9/10

1 CHANGING POPULATION

This core theme provides a background to the key global issues of our time, such as population dynamics, climate change and resource consumption. Four key concepts influence these issues: places, power, processes and possibilities. There are positive aspects of change as well as negative ones. It is necessary to accept responsibility for the causes, and to seek solutions and manage the issues.

You should be able to show:

✔ how population varies between **places;**

✔ the **processes** of population change and how these affect people and places;

✔ that there are population **possibilities** and **power** over the decision-making process.

1.1 POPULATION AND ECONOMIC DEVELOPMENT PATTERNS

- **Population density** – the number of people living in a given area, usually expressed as people per square kilometre. It is calculated by dividing the population of a region by its area.

- **Population distribution** – the location of people within an area. Population is unevenly distributed for a number of reasons. Factors that attract people include mineral resources, temperate climate, the availability of water and fertile, flat land. Factors that repel people include dense vegetation, limited accessibility and political or religious oppression.

- **Voluntary internal migration** – refers to the movement of population away from their home, from one part of a country to another. It occurs when people are free to move where they choose.

- **Core–periphery** – a more-developed part (core) of a country or the world, and a less-developed part (periphery) of a country or the world.

- **Megacity** – a city with more than 10 million inhabitants.

You should be able to show how population varies between places:

✔ Physical and human factors affecting population distribution at the global scale;

✔ Global patterns and classification of economic development:

  ✔ Low-income countries;

  ✔ Middle-income countries and emerging economies;

  ✔ High-income countries;

✔ Population distribution and economic development at the national scale, including voluntary internal migration, core–periphery patterns and megacity growth;

  ✔ Two detailed and contrasting examples of uneven population distribution.

Physical and human factors affecting population distribution at the global scale

Population distribution and density refer to where people live and how many live in a given area. They are both affected by a number of physical and human factors.

 Content link

Physical factors affecting population distribution and density are discussed in options A.3, B.3 and C.1.

Test yourself

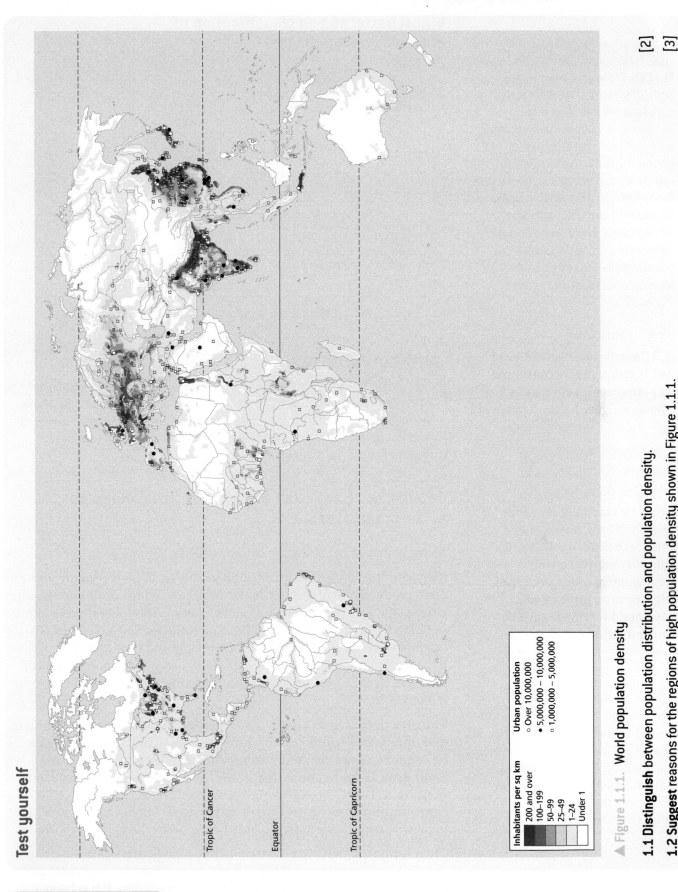

[2]

[3]

▲ Figure 1.1.1. World population density

1.1 Distinguish between population distribution and population density.

1.2 Suggest reasons for the regions of high population density shown in Figure 1.1.1.

Legend (from map):

Inhabitants per sq km
- 200 and over
- 100–199
- 50–99
- 25–49
- 1–24
- Under 1

Urban population
- ○ Over 10,000,000
- ● 5,000,000 – 10,000,000
- □ 1,000,000 – 5,000,000

Tropic of Cancer

Equator

Tropic of Capricorn

>> **Assessment tip**

Population varies between **places.** The distribution of population is influenced by physical factors in the first instance, and increasingly by human factors.

Concept link

PLACES: Places have similarities and differences in terms of their population distribution and economic development. Underlying each of these, there are a range of physical and human factors that ensure each process is dynamic and there is an interrelationship between these two factors. Not only are there internal interactions occurring between places within a country, but also between countries due to regional and global processes.

Test yourself

1.3 Describe the distribution of
(a) high-income countries, and
(b) low-income countries as
shown in figure 1.1.2.   [2+2]

≫ Assessment tip

It is easy to talk about rich and poor countries. However, the World Bank uses a four-fold classification to illustrate the diversity of economic development within countries: high income, upper-middle income, lower-middle income and low income.

Content link

Connect this information with the role of powerful organizations and global groups in helping countries develop, discussed in unit 4.1. Unit 4.2 discusses how global networks and flows can also affect this development.

Content link

The effect of free trade zones (FTZs) on global interactions is discussed in unit 4.3.

Global patterns and classification of economic development

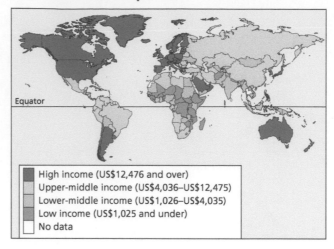

High income (US$12,476 and over)
Upper-middle income (US$4,036–US$12,475)
Lower-middle income (US$1,026–US$4,035)
Low income (US$1,025 and under)
No data

▲ Figure 1.1.2. The World Bank economic classification of countries

≫ Assessment tip

When describing patterns, try to structure your answers so that you show the main features and anomalies (exceptions), and support your answers with examples. So for this question you would actually name countries that are anomalies, for example.

Population distribution and economic development at the national scale

Megacity growth in the Greater Bay Area, China

Megacity clusters will deliver China's future economic growth, and the most productive will be the Greater Bay Area, which combines the nine cities of the Pearl River Delta with the Special Administrative Regions of Hong Kong and Macau. The Area accounts for just 1% of China's land mass, contains nearly 70 million people and produces 37% of the country's exports and 12% of its GDP. The Guangdong province accounts for 22% of China's high-tech exports and this could rise to 40% by 2025.

The growth of megacities can lead to urban sprawl, slum development and income inequalities, which can cause social and political tension. In addition, air quality and water quality is poor, and much of the Pearl River Delta has been degraded. Large cities without affordable housing and efficient public transport can push the poor to live far from jobs and markets, forcing them to choose between long and expensive commutes or living in slums in which they have few rights.

The Pearl River Delta has developed into a large manufacturing region due to a large amount of cheap labour, an excellent sea port, and the development of the Free Economic Zone at Shenzhen. It has been a major attraction for migrants ready to escape rural poverty for the prospects of better-paid work in urban areas. However, many migrants lack the right to education and healthcare in urban areas, and they remain relatively impoverished compared to those urbanites who have rights.

Migration is also affecting rural areas. Most of the wealth in rural areas come from money sent back (remittances) from migrants working in urban areas.

Core–periphery patterns

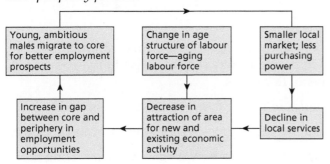

▲ Figure 1.1.4. A model of labour migration and core–periphery inequalities

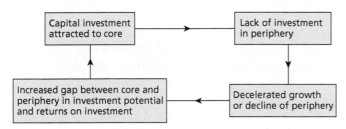

▲ Figure 1.1.5. A model of investment and core–periphery inequalities

Uneven population distribution: China

China's population is concentrated in the eastern part of the country, especially in coastal zones and the lower reaches of river valleys. Much of the rest of the country is characterized by desert (for example, the Gobi Desert), the steep slopes of the Himalayas and the dry grasslands of the north-west.

The uneven population distribution in China results primarily from the country's physical geography. Only a small proportion of the country can provide for rain-fed agriculture—most of the land is too dry or too steep to allow for much agriculture. In addition, the coastal and river locations are the more favoured sites for trade and commerce.

Uneven population distribution: South Africa

The distribution of South Africa's population is very uneven. Some parts of the core economic regions, such as Gauteng province, have population densities of over 1,000 people per square kilometre, whereas large areas of the Northern Cape Province have densities of under five people/per square kilometre. High population densities are found in areas where there are good mineral resources, such as gold and diamonds, good farming potential, and good trading potential, such as Durban and Cape Town.

In general, the population decreases from the south-east to the north-west. This partly reflects the distribution of rainfall in South Africa: the lowest densities are found in the most arid areas and in parts of the mountain regions.

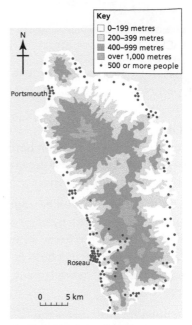

Key
- ☐ 0–199 metres
- 200–399 metres
- 400–999 metres
- over 1,000 metres
- • 500 or more people

Portsmouth

Roseau

0   5 km

▲ Figure 1.1.3. Population distribution in Dominica

Test yourself

1.4 Describe the distribution of population on the island of Dominica. [2]

1.5 Suggest reasons for the distribution of population in Dominica. [3]

Test yourself

1.6 Suggest why some cities develop into megacities. [3]

1.7 Outline the advantages and disadvantages of megacity growth. [3+3]

1.8 Using the data on the Greater Bay Area megacity growth and figures 1.1.4 and 1.1.5, **explain** how megacities can lead to core–periphery inequalities in a country or region. [2+2]

1.9 Using an atlas, **suggest** how physical and human factors have influenced the distribution of population in China. [3+3]

1.2 CHANGING POPULATIONS AND PLACES

• **Natural increase** – the growth in population as a result of birth rates exceeding death rates.

• **Total fertility rate** – the average number of children born to a woman if she lives to the end of her child-bearing years.

• **Life expectancy** – the average number of years to be lived by a group of people born in the same year. Life expectancy at birth is also a measure of overall quality of life in a country.

• **Population structure** – the composition or make-up of the population, for example, age, sex, occupation, race, ethnicity, religion, class. The most commonly used indicators are age and sex, and these are shown using a population pyramid.

• **Dependency ratio** – a measure of the non-workers divided by the workers. It is commonly expressed as the population aged 0–15 (the children) and the population aged 65 and over (the retired) divided by the population aged 16–64 (the workers).

• **Demographic transition** – changes in birth and death rates over time; both rates change from high to low.

• **Forced migration** – migration that occurs due to war, religious persecution, famine, slavery, politics or natural disasters, creating refugees and internally displaced people.

Test yourself

1.10 Define the terms "birth rate" and "death rate".   [1+1]

1.11 Compare the demographic characteristics of Ethiopia with those of China.   [2+2]

1.12 Suggest why population growth varies between China and Ethiopia.   [2]

You should be able to examine processes of population change and their effect on people and places:

✔ Population change and demographic transition over time, including natural increase, fertility rate, life expectancy, population structure and dependency ratios;

   ✔ Detailed examples of two or more contrasting countries;

✔ The consequences of megacity growth for individuals and societies;

   ✔ One case study of a contemporary megacity experiencing rapid growth;

✔ The causes and consequences of forced migration and internal displacement;

   ✔ Detailed examples of two or more forced movements, to include environmental and political push factors, and consequences for people and places.

Population change and demographic transition

▼ Table 1.2.1. Population characteristics for China and Ethiopia

	China	Ethiopia
Birth rate (‰)	12.3	36.5
Death rate (‰)	7.8	7.7
Population aged 0–15 years (%)	17.2	43.5
Population aged 16–65 years (%)	72.0	53.6
Population aged over 65 years (%)	10.8	2.9
Dependency ratio (%)	38.8	86.6

Source of data: Adapted from *CIA World Factbook*

The Demographic Transition Model (DTM) suggests that changes in birth and death rates happen in five stages:

1. High birth rate, fluctuating but high death rate.

2. Birth rate stays high, death rate starts to fall.

3. Birth rate starts to fall, death rate continues falling.

4. Birth rate is low, death rate is low.

5. The birth rate is low and the death rate increases.

▼ Table 1.2.2. Birth and death rate for Bangladesh, 1901–2018

Period	Birth rate (per thousand)	Death rate (per thousand)
1901–11	53.8	45.6
1911–21	52.9	47.3
1921–31	50.4	41.7
1931–41	52.7	37.8
1941–51	49.4	40.7
1951–61	51.3	29.7
1961–74	48.3	19.4
1971–80	47.0	17.2

Period	Birth rate (per thousand)	Death rate (per thousand)
1986	38.9	11.9
1989	–	11.4
1994	27.8	8.6
1998	19.9	4.8
2011	22.9	5.7
2018	18.8	5.4

Dashes indicate no data is available

Source of data: UN Commission on Population and Development and CIA World Factbook

Test yourself

1.13 (a) Using table 1.2.2, **determine** the natural increase for Bangladesh for the period 1901 to 2018. [1]

(b) Identify the period when natural increase was greatest. [1]

(c) Describe the trends in natural increase in Bangladesh between 1901 and 2018. [2]

1.14 Draw an appropriate chart to show the changes in the birth rate and death rate for Bangladesh for the period 1901–2018. [3]

1.15 Describe the main changes you have shown. [3]

1.16 To what extent does Bangladesh follow the Demographic Transition Model (DTM)? [3]

> ### Concept link
>
> **PROCESSES:** Economic, environmental, political and social processes create change within a country's population by influencing fertility, deaths and migration. Changes occur in rural and urban areas and the interactions between these places create further changes, such as the forced movement of people from one place to another. The process of internal displacement of people can be linked to environmental and political change, and this migration can subsequently create pressure elsewhere within a country.

The consequences of megacity growth for individuals and societies

Case study: A contemporary megacity experiencing rapid growth—the Greater Bay Area

China has completed building a 55 km bridge connecting the former European colonies of Hong Kong and Macau with the city of Zhuhai. It was designed by the Chinese government to connect these two semi-autonomous regions more closely to the mainland, both economically and politically. In addition, a new US$11 billion rail link, which links Hong Kong into China's vast high-speed rail network, is a crucial element in Beijing's plan to integrate Hong Kong and Macau with nine neighbouring urban areas, including the megacities of Shenzhen and Guangzhou.

The Greater Bay project covers an area containing nearly 70 million people with a US$1.5 trillion economy, larger than some G20 countries such as Australia, Indonesia and Mexico. The area's economy is predicted to nearly double to US$2.8 trillion by 2025. Guangdong is the "workshop of the world", exporting US$670 billion of goods in 2017. Its economy is more driven by private enterprise than any other area of China. The Greater Bay Area has three of the world's 10 busiest container ports—in Hong Kong, Guangzhou and Shenzhen—and thriving international airports in all three cities.

The Greater Bay Area plans to promote large-scale business opportunities, but many Hong Kong residents claim that they have lost freedoms and autonomy. Critics argue

> ### Content link
> The factors affecting the economic activities of urban environments as megacities are discussed further in option G.1.

▲ Figure 1.2.1. The Hong Kong-Zhuhai-Macau bridge took over 15 years to build and cost nearly US$20 billion

Test yourself

1.17 State one advantage and two disadvantages of high population density. [1+2]

1.18 Suggest why the Greater Bay Area continues to attract many migrants. [3]

1.19 Outline the potential impacts of megacity growth for individuals and societies. [3]

>> Assessment tip

Remember to read the question carefully, and then answer it. Sometimes, as here, you are asked to give different numbers of advantages/disadvantages or social/economic/environmental reasons. In question 1.17, you will only receive credit for one advantage, but you are expected to give two disadvantages.

Test yourself

1.20 Outline the main areas of the world in which there are displacements due to conflict and violence. [3]

 Content link

The displacement of populations and the resulting influence on cultural diversity and identity is discussed further in unit 5.2.

that one of the most visible signs of Beijing's political intentions is the creation of an additional border crossing facility in the new rail station in West Kowloon, Hong Kong.

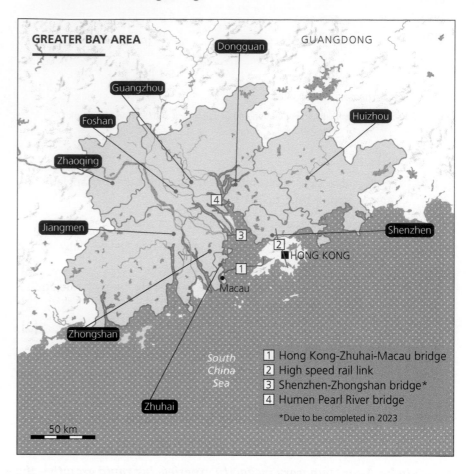

▲ Figure 1.2.2. The Greater Bay Area megacity region

The causes and consequences of forced migration and internal displacement

Causes of forced movement and internal displacement include civil war, political uprising, the rise of terrorism and tribal conflict. Environmental causes include flooding, drought, hurricanes, global climate change and volcanic eruptions.

The Democratic Republic of Congo (DRC) has over 2.2 million internally displaced people (IDPs), mainly due to violence and years of war and disaster. Inter-communal fighting and conflicts between tribal groups and the country's armed forces have led to many IDPs, especially in North Kivu and South Kivu. In contrast, in the Horn of Africa (Somalia and Ethiopia), years of drought have led to crop failure, livestock deaths, food insecurity and malnutrition. In addition, conflict and violence have forced many people to become IDPs, and some have been displaced due to heavy "taxation" by the non-state armed group al-Shabaab.

In 2016, flooding in the Yangtze River Basin in China displaced over 7 million people. Also in that year, nearly 10% of Cuba's population were evacuated ahead of Hurricane Matthew, while in Haiti around 550 people lost their lives, half a million people were displaced and up to 90% of homes were destroyed in the worst affected areas.

1.3 CHALLENGES AND OPPORTUNITIES

You should be able to assess population possibilities and power over the decision-making process:

✔ Global and regional/continental trends in family size, sex ratios and aging/greying;

✔ Policies associated with managing population change, focusing on:

  ✔ Policies related to aging societies;

  ✔ Pro-natalist or anti-natalist policies;

  ✔ Gender equality policies and anti-trafficking policies;

✔ The demographic dividend and the ways in which population could be considered a resource when contemplating possible futures;

  ✔ One case study of a country benefiting from a demographic dividend.

Global and regional/continental trends in family size, sex ratios and aging

The highest fertility rates (more than five children per woman) are found in Central and West Africa, whereas the lowest rates (fewer than two children per woman) are found in North America and Europe. The most rapid decline in the fertility rate has been in North Africa, where the rate fell from about seven in the 1950s to around three in the 2010s. However, globally, fertility rates are declining around the world.

As a result of the falling fertility rates, family size around the world is generally decreasing. Reasons for this include more working women, the high cost of having children, the high cost of housing, longer working hours, marital breakdown, greater availability of contraception and/or the desire for a more materialistic and individualistic lifestyle.

Sex ratios vary considerably around the world. The UAE has the most men (274.0) for every 100 women. Relatively high ratios of men to women are found across the Middle East and North Africa, China, India and western Asia. In contrast, Mauritius has the fewest men (84.5) for every 100 women. Low ratios of men to women are found across Russia, parts of eastern Europe and Japan.

Policies associated with managing population change

Policies related to aging societies

According to the World Health Organization:

• between 2015 and 2050, the proportion of the world's population aged over 60 years will double from 12% to 24%

• by 2020 the number of people aged over 60 years will outnumber children aged under 5 years

• by 2050, 80% of older people will be living in low- and middle-income countries.

• **Sex ratio** – the ratio of males to females in a population or cohort (age group).

• **Aging population** – a situation in many HICs where the average age of the population is increasing. It occurs when birth rates fall and the number of elderly people increases. Countries with older populations (a high percentage aged 65 and over) need to invest more in the health sector and in pensions. This may create problems if the workforce is small or reducing.

• **Pro-natalist policies** – population policies that encourage families to have more children, for example longer maternity/paternity leave, increased child allowance.

• **Anti-natalist policies** – population policies that encourage families to have fewer children, for example forced abortions, sterilization.

• **Gender equality** – treating males and females the same.

• **Trafficking** – taking people against their will and forcing them into occupations they do not wish to do.

• **Demographic dividend** – the benefits that come from a very large proportion of the population being in the adult age range, and a reduced proportion of young and/or elderly.

Test yourself

1.21 Suggest reasons for variations in gender differences around the world. [3]

>> **Assessment tip**

A common misconception is that aging populations are a drain on societies. Some aging populations can be a drain, but many are well-off and provide valuable social and economic services such as childcare and charity work, and they may spend a lot on travel and recreation.

>> **Assessment tip**

When describing data from a graph or table, look for the maximum and minimum values and any trends, and point out exceptions (anomalies). Remember to use the data (manipulate it or transform it) rather than just copying it.

Test yourself

1.22 Outline the opportunities and constraints of having
(a) an aging population, and
(b) a youthful population.   [3+3]

Concept link   ⌬

POSSIBILITIES AND POWER:
A country's population will be influenced by political decisions that focus on different demographic sections in a country. For example, policies such as implementing pro-natalist and anti-natalist policies, or ensuring that an emerging youthful working population will find employment opportunities, may help to manage change and help a country develop in a sustainable manner. Governments can be effective in creating change, such as China's One Child Policy, although there are many other governments in other countries that have struggled to influence their citizens with regard to natalism. Contemporary issues such as human trafficking not only relate to the power of authorities within a country's borders, but also cross-border cooperation in order to manage the illegal movement.

With aging, there is a gradual decrease in physical and mental capacity, a growing risk of disease, and ultimately death. However, there are other life changes, such as retirement, relocation to more appropriate housing, and the death of partners, friends and relatives.

Policies to deal with aging societies must therefore:

- commit to healthy aging, that is, improve the health of the elderly
- align health systems with the needs of the older population
- develop systems to deal with long-term care
- provide pensions and financial assistance.

However, these things require a great deal of funding which may increase the burden on taxpayers, for example, increased taxes and/or raising the age of retirement.

Pro-natalist and anti-natalist policies

Pro-natalist policies are designed to encourage families to have more children, while anti-natalist policies attempt to reduce the birth rate. Singapore has had both anti-natalist policies and pro-natalist policies.

In 1960, Singapore had a total fertility rate of over 5.5. This fell to under 1.5 by 1985, and has remained low since then.

Anti-natalist policies in Singapore (1972–87) included:

- creation of the Family Planning and Population Board
- increased access to family planning clinics
- promotion of sterilization programmes
- increased access to low-cost contraception
- use of the media to promote smaller families
- free education and low-cost healthcare for smaller families.

However, from 1987, pro-natalist policies encouraged families to have three or more children if they could afford to.

- The Family Planning and Population Board was abolished.
- Child benefits were increased, especially for those with higher academic qualifications.
- Maternity leave was increased.
- There was government-sponsored childcare.
- Sterilization and abortion were discouraged.
- State-sponsored dating agencies were established.

Overall, there were slight increases in the fertility rate when these measures were introduced but they were short-lived. Some businesses were against the increased maternity leave, and some individuals felt their "free choice" was being eroded.

Gender equality policies

According to the UN:

- in Sub-Saharan Africa and Western Asia girls face barriers to entering primary and secondary schools
- women in North Africa account for less than 20% of jobs in the non-agricultural sector
- on average, women in the labour market earn 24% less than men globally.

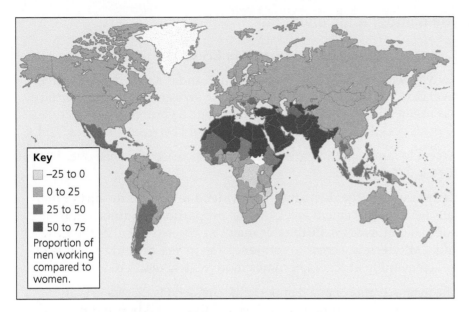

Content link
Unit 5.1 discusses UN policies for empowering women and indexes for measuring gender equality in more detail.

▲ Figure 1.3.1. Inequalities in employment (the difference between the proportion of men and women working)

Source of data: International Labour Organization

The UN Sustainable Development Goals (SDG) aim to:

- end all forms of discrimination against women and girls everywhere
- eliminate all forms of violence against women and girls
- recognize and value all forms of unpaid care and domestic work
- give women equal rights to economic resources, ownership and control of land and property, financial services, inheritance and natural resources.

Anti-trafficking policies

Human trafficking is a crime that strips people of their rights, dignity and hopes. It is a global problem.

▼ Table 1.3.1. Modern-day slavery: victims and profits by region

	Developed economies and the EU	Latin America and the Caribbean	Central and Southern Europe and the CIS	Africa	Middle East	Asia-Pacific
Victims	1.5 m	1.8 m	1.6 m	3.7 m	600,000	11.7 m
Cost (US$)	$44.9 bn	$12 bn	$18 bn	$13 bn	$8.5 bn	$52 bn
Cost per victim	$34,800	$7,500	$12,900	$3,900	$15,000	$5000

Source of data: International Labour Organization (2014)

There are a number of organizations and charities raising awareness of human trafficking.

The Blue Heart Campaign (www.unodc.org/blueheart/) raises awareness of the impact of trafficking and the need to fight it; it also attempts to produce action to stop trafficking. To raise awareness, it encourages people to wear the Blue Heart to show solidarity with the victims of trafficking. (The Blue Heart is a symbol for the sadness of those trafficked and the cold-heartedness of the traffickers.)

The United Nations Voluntary Trust Fund for Victims of Trafficking in Persons, Especially Women and Children was created in 2010 as an integral part of a global effort to address trafficking in persons. The Trust Fund provides humanitarian, legal and financial aid to victims of trafficking. It supports NGOs that help people who have been exploited by trafficking.

Test yourself

1.23 Identify the region(s) where inequality in employment is greatest. [2]

1.24 Evaluate the type of map shown in Figure 1.3.1 (choropleth—density shading) and its scale as a means of showing variations in employment. [3]

Content link
Unit 4.2 discusses illegal flows, such as human trafficking, further.

Test yourself

1.25 Describe the number of victims and profits by geographic region, based on the data in table 1.3.1. [3]

1.26 Briefly **explain** the reasons for the use of the Blue Heart as a symbol of trafficking. [2]

The demographic dividend

Case study: Demographic dividend in Kenya

Kenya's fertility rate is still relatively high and is above the replacement level (2.1), but it fell by over 50% between 1978 and 2014. High fertility has resulted in rapid population growth and a youthful age structure. Kenya's population increased almost fourfold between 1969 and 2014.

Access to family planning is projected to increase from just over 50% in 2014 to 70% by 2030.

Kenya can expect a demographic dividend to occur if the birth rate falls below replacement level and the proportion of adults in the population increases. Birth rates may be expected to fall as the IMR and CMR decline, access to contraception increases, and the value of girls remaining at school becomes more widely realized.

The main advantage of a demographic dividend is that there is an increase in the proportion of workers relative to children and the elderly. Thus, the dependency ratio decreases. Having a larger workforce should result in greater economic output and greater revenue for the country. However, if there are insufficient jobs for the workforce, having too many adults could lead to unemployment, underemployment and political and social tensions. Many people may have to leave the country in search of employment, and that could lead to families being separated.

QUESTION PRACTICE

On the right, a Lorenz curve shows the inequality in the distribution of population in China.

a) Using the Lorenz curve:

   i) **state** the proportion of people living on the most densely populated 10% of the land area in China [1]

   and

   ii) **state** the proportion of people living on the least densely populated 10% of the land area in China. [1]

b) **Suggest** two physical factors that can lead to uneven population distribution. [2 + 2]

c) Using an example that you have studied, **explain** the impact of internal migration on the source regions. [4]

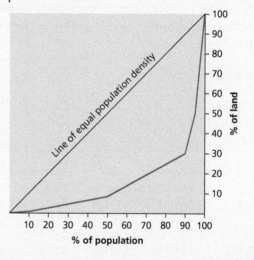

Essay

"Of all the possible challenges facing different countries, demographic issues are of greatest concern."
To what extent do you agree with this statement? [10]

How do I approach these questions?

a) This question requires you to read off the graph.

   i) The reading for the most densely populated area should be straightforward.

   ii) The reading for the least densely populated area requires some data manipulation. To read off the Lorenz curve for the least densely populated 10% of the land, this must be taken away from 100% (that is, take the reading for 90% of the area, and then the value for the percentage of population living on 90% of the land must be taken away from 100% of the population).

b) You are asked to suggest (identify) two physical factors and to suggest how they affect distribution. One mark will be awarded for the identification of a valid physical factor and another mark for further development/exemplification. The explanation will need to be developed (more detailed and/or have a supporting example) in order to get the second mark. There is no credit for the identification/explanation of any human factors.

c) In this question the impacts have to be related to the source (where the migrants come from). No credit is given for impacts on the destination. Each of the impacts needs to be developed for full marks, that is, a more detailed explanation of the impacts in the source area.

Essay

A well-structured answer is needed for this extended response question. Half of your essay should be an examination of either aging populations (Stage 5 of the demographic transition model, DTM) or youthful populations, and the consequences related to their respective populations. Problems may relate to health, education or the cost of services needed, while opportunities may be economic and/or social. The other half of your essay should counter the argument in the question, potentially by covering other challenges to countries such as migration and climate change. Your essay should include appropriate terminology, located examples and supporting facts and figures.

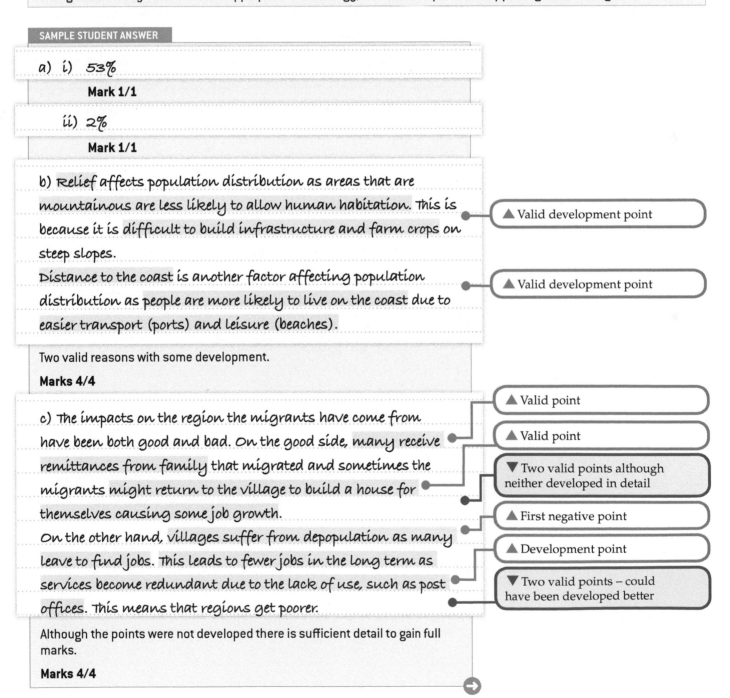

SAMPLE STUDENT ANSWER

a) i) 53%

Mark 1/1

ii) 2%

Mark 1/1

b) Relief affects population distribution as areas that are mountainous are less likely to allow human habitation. This is because it is difficult to build infrastructure and farm crops on steep slopes.

▲ Valid development point

Distance to the coast is another factor affecting population distribution as people are more likely to live on the coast due to easier transport (ports) and leisure (beaches).

▲ Valid development point

Two valid reasons with some development.

Marks 4/4

c) The impacts on the region the migrants have come from have been both good and bad. On the good side, many receive remittances from family that migrated and sometimes the migrants might return to the village to build a house for themselves causing some job growth.
On the other hand, villages suffer from depopulation as many leave to find jobs. This leads to fewer jobs in the long term as services become redundant due to the lack of use, such as post offices. This means that regions get poorer.

▲ Valid point

▲ Valid point

▼ Two valid points although neither developed in detail

▲ First negative point

▲ Development point

▼ Two valid points – could have been developed better

Although the points were not developed there is sufficient detail to gain full marks.

Marks 4/4

Essay

"Of all the possible challenges facing different countries, demographic issues are of greatest concern." **To what extent** do you agree with this statement?

There are many challenges in different countries. These include demographic, social, economic, political and environmental.

▲ Clear introduction—suggests the likely structure of the essay

Often the challenges are a mixture of more than one e.g. too many people (e.g. in megacities) can lead to unemployment (economic challenge) which in turn leads to poverty (social challenge).

▲ Valid challenge and nominal support

For many countries with an ageing population e.g. Japan or China, there are problems with a shrinking workforce. Germany

▲ Further development/ exemplification

accepted some one million Syrian migrants in order to make up the short-fall of its declining workforce. Other problems related to an ageing population can be the high cost of pensions and health

▲ Generic challenges of an ageing population

care, and the need for sheltered accommodation for the elderly.

▲ Generic challenges of a youthful population

Equally, there are problems for youthful populations – pressure on school places, clinics, not enough jobs and so on. Many LICs, such as Niger and Mali, have youthful populations but high

▲ Some development and support

rates of unemployment.

▲ Start of the counter-argument

In other countries, the problems do not necessarily stem from demographic challenges. For people living in low-lying islands, rising sea levels, increasing storms and contamination

▲ Valid point—climate change refugees

of freshwater sources are a bigger concern e.g. in Kiribati.

For others, e.g. in Sudan and South Sudan, desertification and falling food yields are a more pressing concern.

▲ Second valid point

This has an impact on population, although arguably has been caused by increased population pressure on the environment. In Yemen, the major concerns are civil war and the war with Saudi Arabia. The air strikes (bombings) by Saudi Arabia and the blockade of Yemen's ports are major concerns, and up to 12 million people are at risk of starvation.

In other rich countries, other concerns may be more pressing. In the UK, the Brexit negotiations have created major uncertainty regarding the countries' economic and political future. Moreover, it is creating uncertainty for non-UK nationals, whose future in the UK is in question. For countries such as China, trade wars with the USA are causing major economic challenges. The country has other challenges, such as pollution and a declining labour force, although the latter could be partially solved by speeding up rural-urban migration.

So, overall, demographic challenges are important but not necessarily the greatest challenge in all countries. Nevertheless, people are affected by other challenges, so there is an impact on the quality of life for people.

▲ Population as a potential cause and impact

▲ Another valid example—good details

▲ Contemporary example—good point

▲ Demographic challenge of Brexit

▲ Another contemporary challenge—identifies a number of challenges in China

Good account—focused on the question; has good supporting evidence. Critical thinking/evaluation is implied but not explicit.

Marks 9/10

2 GLOBAL CLIMATE—VULNERABILITY AND RESILIENCE

Four key concepts influence these issues: places, power, processes and possibilities. There are positive aspects of change, as well as negative ones. It is necessary to accept responsibility for the causes, and to seek solutions and manage the issues.

You should be able to show:

✔ how natural and human **processes** affect the global energy balance;

✔ how the effects of global climate change on **places**, societies and environmental systems;

✔ that there are **possibilities** for responding to climate change and **power** over the decision-making process.

2.1 THE CAUSES OF GLOBAL CLIMATE CHANGE

- **Albedo** – the amount of incoming solar energy reflected back into the atmosphere by the Earth's surface.

- **Anthropogenic** – human-related processes and/or impacts.

- **Energy balance** – the balance between incoming short-wave radiation and outgoing short-wave and long-wave radiation.

- **Enhanced greenhouse effect** – the increasing amount of greenhouses gases in the atmosphere, as a result of human activities, and their impact on atmospheric systems including global warming.

- **Global warming** – the increase in temperatures around the world that has been noticed since the 1960s, and in particular since the 1980s.

- **Greenhouse effect** – also called the natural greenhouse effect, this is the process by which certain gases (water vapour, carbon dioxide, methane

You should be able to show how natural and human processes affect the global energy balance:

✔ The atmospheric system, including the natural greenhouse effect and energy balance;

✔ Changes in the global energy balance, and the role of feedback loops;

✔ The enhanced greenhouse effect and international variations in greenhouse gas sources and emissions, in relation to economic development, globalization and trade.

The atmospheric system

The natural greenhouse effect and the enhanced greenhouse effect (global warming)

The natural greenhouse effect is the process by which certain gases (greenhouse gases) allow short-wave radiation from the Sun to pass through the atmosphere but trap an increasing proportion of outgoing long-wave radiation from the Earth. This radiation leads to a warming of the atmosphere. The greenhouse effect is a good thing, for without it there would be no life on Earth. For example, the Moon is an airless planet that is almost the same distance from the Sun as the Earth. However, daytime temperatures on the Moon may reach as high as 100°C, whereas by night they may be −150°C. Average temperatures on the Moon are about −18°C compared with about 15°C on Earth. The Earth's atmosphere therefore raises temperatures by about 33°C.

There are a number of greenhouse gases. Water vapour accounts for about 95% of greenhouse gases by volume and for about 50% of the greenhouse effect. However, the gases mainly implicated in global warming are carbon dioxide, methane and chlorofluorocarbons.

Carbon dioxide (CO_2) levels have risen from about 315 parts per million (ppm) in 1950 to over 400 ppm in 2015, and are expected to reach 600 ppm by 2050. The increase is due to human activities: burning fossil fuels (coal, oil and natural gas) and land-use changes such as deforestation. Deforestation of the tropical rainforest is a double blow, since it not only increases atmospheric CO_2 levels but it also removes the trees that convert CO_2 into oxygen. Carbon dioxide accounts for about 20% of the greenhouse effect but an increased proportion of the enhanced greenhouse effect.

Methane is the second-largest contributor to global warming, and its presence in the atmosphere is increasing at a rate of 1% per annum. It is estimated that cattle convert up to 10% of the food they eat into methane and emit 100 million tonnes of methane into the atmosphere each year. Natural wetland and paddy fields are other important sources: paddy fields emit up to 150 million tonnes of methane annually, while, as global warming increases, bogs trapped in permafrost will melt and release vast quantities of methane.

The enhanced greenhouse effect is the impact of increasing levels of greenhouses gases in the atmosphere as a result of human activities. It is often referred to as global warming. Global climate change refers to the changes in the global patterns of rainfall and temperature, sea level, habitats and the incidence of drought, floods and storms, resulting from changes in the Earth's atmosphere, believed to be caused mainly by the enhanced greenhouse effect.

The increase in the world's greenhouse gases is linked to industrialization, trade and globalization. As industrialization has increased, so too has atmospheric CO_2. Many LICs and NICs are actively industrializing and adopting a consumer culture. Industrial activity among the NICs has the potential to add to atmospheric CO_2. Nevertheless, the per-capita emissions in HICs are responsible for much of the growth in atmospheric CO_2.

Changes in the global energy balance

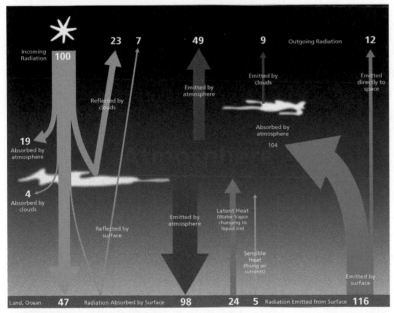

▲ Figure 2.1.1. The Earth's energy balance

and chlorofluorocarbons (CFCs)] allow short-wave radiation from the Sun to pass through the atmosphere and heat up the Earth, but trap a proportion of long-wave radiation from the Earth. This radiation leads to a warming of the atmosphere.

- **Feedback** – the ways that changes in an environment may be accelerated or modified by the processes operating in a system.

- **Positive feedback** – changes in a system that lead to greater deviation from the original condition (also known as cumulative causation or a vicious circle).

- **Negative feedback** – changes in a system that occur and lead to other changes, but eventually the whole system stabilizes.

Test yourself

2.1 According to figure 2.1.1, **determine** the amount of the incoming solar radiation that is absorbed by the Earth's surface. [1]

2.2 Outline the ways in which solar radiation differs from the Earth's radiation. [2]

2.3 Compare the incoming sources of energy in the atmosphere with the outgoing energy sources. [2]

2.4 Explain two processes in which global energy is balanced. [2+2]

>> **Assessment tip**

Do not confuse the ozone layer with the greenhouse effect—they are very different. The ozone layer protects the Earth from harmful ultraviolet radiation whereas the greenhouse effect is responsible for raising the temperature on Earth and making life possible. (Ozone is a minor greenhouse gas, but its contribution to the greenhouse effect is very small.)

>> **Assessment tip**

Be very clear about the difference between the *natural* greenhouse effect (which is a good thing and vital for life) and the *enhanced* greenhouse effect (which is related to human activities and is not so good for everyone).

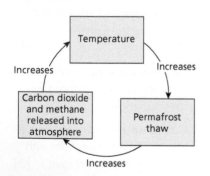

▲ **Figure 2.1.2.** A positive feedback mechanism involving methane and enhancing climate change

Solar radiation variations

There have been many variations in solar radiation, and these have caused changes in the Earth's climate over the geological past. However, variations in solar radiation have not been significant in recent warming. Since the 1970s, fluctuations in solar radiation have been measured by satellites. Solar activity has been declining since the 1960s. In addition, there has been no correlation between sunspot activity and global warming. However, some of the variation in solar radiation reaching the Earth's surface can be put down to periodic volcanic activity, such as the eruption of Mt Pinatubo in the Philippines in 1991, and human-caused pollution, causing global dimming.

Feedback loops

▼ Table 2.1.1. Some albedo values for terrestrial surfaces

Surface	Albedo (%)
Water (Sun's angle over 40°)	2–4
Water (Sun's angle less than 40°)	6–80
Fresh snow	75–90
Old snow	40–70
Dry sand	35–45
Dark, wet soil	5–15
Dry concrete	17–27
Black road surface	5–10
Grass	20–30
Deciduous forest	10–20
Coniferous forest	5–15
Crops	15–25
Tundra	15–20

Source of data: Adapted from Barry, R. and Chorley, R., *Atmosphere, weather and climate*, Routledge (1998)

The enhanced greenhouse effect and international variations in greenhouse gas sources and emissions

In 1990, total CO_2 emissions were about 22 billion tonnes. Europe and Eurasia produced most of the CO_2 emissions, about 8 billion tonnes, followed by North America, with about 6 billion tonnes. China only produced about 2 billion tonnes, and the rest of Asia was similar. By 2016 the Asia-Pacific region produced almost half of global emissions, up from 25% in 1990. China alone produced about 10 billion tonnes of CO_2, Europe and Eurasia's output had fallen to around 6 billion tonnes, and North America had remained steady overall.

2.2 THE CONSEQUENCES OF GLOBAL CLIMATE CHANGE

You should be able to show the effects of global climate change on places, societies and environmental systems:

✔ Climate change and the hydrosphere, atmosphere and biosphere;

✔ Impacts of climate change on people and places, including health hazards, migration and ocean transport routes.

Climate change and the hydrosphere, atmosphere and biosphere

Water stored in ice and oceans, and changing sea levels

Europe's glaciers lost around 25% of their mass between 2006 and 2014.

- **Sea levels are expected to rise by one metre by 2100.**

- There has been a 13.3% decrease in Arctic sea ice each decade since 1980.

- By 2040, summer sea ice is likely to be limited to the northern coasts of Canada and Greenland.

- By 2080, arctic summer sea ice is expected to disappear completely.

- By 2100, arctic temperatures will be as high as 7°C above pre-industrial levels.

Carbon stored in ice, oceans and the biosphere

Glaciers store carbon derived from primary production on the glacier and deposition of materials such as soot or by-products of the combustion of fossil fuels. Measurements in Greenland and Antarctica suggest that the amount of organic carbon lost from glaciers will increase by 50% between 2015 and 2050. That equates to roughly half of the total amount of carbon carried by the Mississippi River to the ocean each year.

Permafrost contains vast amounts of carbon. When permafrost melts, carbon is released either as carbon dioxide or as methane, stored in frozen form. If permafrost continues to melt, some 190 gigatonnes of carbon could be released into the atmosphere by 2200.

Methane released from seafloor permafrost could also contribute to ocean acidification. Currently about 25% of human-produced carbon dioxide emissions are absorbed by the oceans. Carbon dioxide reacts with seawater to make it more acidic. This change affects many marine organisms.

Incidence and severity of extreme weather events including drought

Trends towards extreme climate change are likely to increase. This includes the weakening of the North Atlantic Drift (NAD) and the meandering behaviour of the jet stream. Increased flows of freshwater from Greenland's ice sheet caused a 30% slowdown of the NAD between 2009 and 2010. A complete switch-off of the NAD could reduce land temperatures in the UK, Greenland, Iceland and Scandinavia by 9 °C.

- **Saltwater intrusion** – the contamination of groundwater by seawater.

- **Biome** – a large-scale naturally occurring ecosystem, identifiable on a global or regional scale.

- **Ecosystem services** – the products and services provided by ecosystems, such as climate regulation, flood regulation, oxygen, food, timber and water.

- **Ecological threshold** – the point at which there is an abrupt change in an ecosystem properties or quality.

- **Resilience** – the ability of a population or a human or natural system to absorb change without having to make a fundamental change.

- **Threshold** (or **tipping points**) – the critical level at which change is irreversible.

- **Vulnerability** – the degree to which a human or natural system is susceptible to, and unable to cope with, the adverse impacts of climate change.

 Content link

Changes in sea levels are also discussed in option B.2.

>> **Assessment tip**

Carbon dioxide (CO_2) and methane (CH_4) have dissimilar names, but both contain carbon, and both are greenhouse gases.

Concept link 🔗

PLACES: All places will be affected by global climate change, with some places experiencing benefits, and others will be negatively impacted. The natural and human worlds are changing, and places at varying levels of development will suffer contrasting impacts. For example, some places may suffer environmental degradation and populations will relocate to another place, while others may experience rising food prices due to a reduction in the agricultural yield in their country, or perhaps elsewhere in the world.

▲ **Figure 2.2.1.** Global climate change is leading to the drying of many soils

🔗 Content link

Future possibilities in sustainable food production are discussed in option F.4.

Due to higher temperatures, there may be an increase in cyclone activity (including hurricanes). Increased evaporation may potentially lead to more frequent and intense storm activity, particularly in coastal areas.

Wildfires have become more common in high latitudes. As firefighting in remote areas is difficult, many of these fires burn for months, adding carbon to the atmosphere.

Spatial changes in biomes, habitats and animal migration patterns

Tropical forests are beginning to die back due to the increased severity and frequency of droughts. In 2005 and 2010, two severe droughts led to the Amazon rainforest emitting more carbon than it absorbed. Although Indonesia's forests are only 20–25% of the size of the Amazon rainforest, forest fires there emit massive amounts of carbon as many of the forests grow on carbon-rich peat.

Climate change can affect where species live, their food supply and the timing of biological events. Projected climate change may increase the extinction of species in sensitive areas. Climate change may affect the capacity of ecosystems to survive extreme events, such as fires, droughts and floods. Mountain and arctic ecosystems are especially vulnerable, as species have fewer places in which to take refuge. Many fish species have migrated to higher latitudes and warmer conditions.

Warming may force species to higher altitudes or latitudes. As sea levels rise, saltwater intrusion into freshwater may adversely affect some species. As rivers warm, warm water fish species are replacing coldwater species such as trout and salmon. The coldwater species are projected to lose around 50% of their habitat by 2080.

The impact of climate change can pass up through the food web. Reductions in sea ice in the Arctic lead to a decline in the algae that are eaten by zooplankton, and in turn by cod. Cod are an important food source for seals, which are fed on by polar bears.

Agricultural crop yields, limit of cultivation and soil erosion

- In some areas, crop yields will reduce due to warmer and drier conditions. In sub-Saharan Africa, an increase of 1.5–2.0°C will lead to a decrease of millet and sorghum areas by 40–80%.

- The limits of cultivation may move further north in North America and Russia due to rising temperatures in the tundra, which will lead to the possibility of agriculture and increasing growing seasons.

- Soil erosion, land degradation and desertification has increased the size of the Sahara Desert during the 20th century.

Impacts of climate change on people and places, including health hazards, migration and ocean transport routes

Certain population groups are more vulnerable to climate change. These include the poor, young, elderly and sick, and people living in vulnerable areas. Low-lying coastal areas are at risk from a variety of threats such as flooding, saltwater intrusion and storm surges. Cities are also vulnerable due to the large concentrations of people there.

Cities increase the risk of heatwaves due to the formation of urban microclimates and the heat-island effect. Many cities have an aging infrastructure, including drainage and sewer systems, flood protection schemes, transport and power supply systems.

Indigenous populations are vulnerable for many reasons. They:

- rely on the natural environment for food and cultural practices, as well as for income

- live in isolated and/or low-income communities

- have high rates of uninsured individuals

- have high rates of existing health risks compared to other groups.

>> **Assessment tip**

It is a common misconception that only the poor will be affected by climate change. Although certain groups are more vulnerable to climate change (for example, poor, indigenous populations and refugees), many middle- and high-income people will also be affected. Food prices will rise, food scarcity will increase, insurance premiums will increase, and the likelihood of water shortages will also increase.

>> **Assessment tip**

Try to get across the complexity of the impact of climate change on human health. For example, people's health will not only be affected by the spread of diseases such as malaria, but many people will be affected by severe dehydration due to prolonged drought, and fatigue due to high temperatures, which may lead to an increase in illness (morbidity) and death rates (mortality).

Content link
Managing climate change in urban environments is discussed in option G.4.

Diseases such as malaria will become more common as temperatures rise. This is because the mosquitoes need temperatures above 20°C in order to breed.

Around 60 million people in the Indus and Brahmaputra river basins rely on glacial meltwater for their water supply. With glaciers melting faster, there is likely to be an increase in the short-term flow of water, but a long-term decrease as the sources of water disappear.

Considering all of the information above, planners are anticipating a large-scale increase in the number of climate-change refugees. In some locations, such as the Pacific island of Kiribati, some people have already been forced to leave their homes due to rising sea levels and saltwater intrusion into freshwater.

Transport routes, such as the north-west passage from the USA to Arctic Canada, may open up, as well as the seas to the north of Siberia. Not only would this allow for year-round transport, it may also facilitate oil exploration.

Test yourself

2.10 Explain how climate change may affect **(a)** agriculture and **(b)** ecosystems. [3+3]

2.3 RESPONDING TO GLOBAL CLIMATE CHANGE

- **Adaptation** – initiatives and measures to reduce the vulnerability of human and natural systems to climate change.

- **Mitigation** – attempts to reduce the causes of climate change.

- **Risk** – the probability of a hazard event causing harmful consequences (expected losses in terms of death, injuries, property damage, economy and environment).

- **Geoengineering** – schemes designed to tackle the effects of climate change directly, usually by removing CO_2 from the air or limiting the amount of sunlight reaching the planet's surface (also known as climate engineering).

- **Stakeholder** – a person, community and/or an organization that has an interest in something, For example, stakeholders in climate change might include farmers, oil companies, residents in low-lying coastal areas and so on.

Test yourself

2.11 Using examples, briefly **explain** how vulnerability varies with **(a)** location, **(b)** wealth, **(c)** gender, **(d)** age, **(e)** education and **(f)** risk perception.
[2+2+2+2+2+2]

Test yourself

2.12 Suggest why London is vulnerable to flooding. [2]

2.13 Suggest why Cape Town is running out of water. [2]

2.14 Outline the measures that have been taken to reduce water consumption in Cape Town. [3]

You should be able to show examples of possibilities for responding to climate change and power over the decision-making process:

✔ Disparities in exposure to climate-change risk and vulnerability, including variations in people's location, wealth, social differences (age, gender and education) and risk perception;

✔ Government-led adaptation and mitigation strategies for global climate change;

✔ Civil society and corporate strategies to address global climate change.

Disparities in exposure to climate-change risk and vulnerability

There are many disparities in exposure to climate change, such as by location, wealth, gender, age, education and risk perception.

London's exposure to climate-change risk and vulnerability

London is already vulnerable to extreme weather, namely floods, droughts, heatwaves and very cold weather. It is likely that in a warming world, London will experience warmer, wetter winters and hotter, drier summers. Very cold winters will still occur, although they will become less frequent. Sea level will continue to rise for centuries.

London is vulnerable to flooding from many sources: tidal flooding from the North Sea, river flooding from the Thames and heavy rainfall events. Some 15% of the city lives on a floodplain. That means that over 1.25 million people, more than 480,000 properties and a great deal of key infrastructure (transport links, schools and hospitals) are at risk of flooding.

Drought is likely to increase in future due to decreased summer rainfall and increased demand for water. Although there were droughts in 2003 and 2006, and floods in 2000 and 2002, London was able to deal with them.

Climate change and Cape Town

Cape Town is running out of water. By January 2018, following three dry years, dam levels of usable water were down to 17% of capacity and the authorities were preparing to shut off residential supplies. This would have left around 4 million residents reliant on standpipes. In June 2018 heavy rains refilled Cape Town's dam to 43% capacity, and Day Zero was put back until sometime in 2019.

Given the huge disparities in South African society, there is plenty of scope for resentment. Some are sharing tips on how to wash in a bucket and reuse the contents. Farms and hotels have halved water use. Others are exceeding the 50 litres per day recommended by the city's authorities, emptying supermarket shelves of bottled water. Thus, water usage is still too high and Cape Town may become the first major city in the developed world that runs out of water.

South Africa's weather services have told politicians that their models no longer work and their long-term climate-change predictions have arrived 10 years early.

Investments that would have failed cost-benefit analyses 10 years ago—expensive desalination plants in Cape Town's case—now look essential.

Government-led adaptation and mitigation strategies for global climate change

Mitigation strategies for global climate change

Mitigation refers to attempts to reduce the causes of climate change. Many of these are shown in table 2.3.1.

National and international methods	Individual methods
• Control the amount of atmospheric pollution • Geoengineering • Develop carbon-capture schemes • Develop renewable energy sources • Set limits on carbon emissions • Ocean fertilization • Carbon-trading schemes • Carbon-offset schemes • Introduce carbon taxes	• Use public transport • Use locally produced foods • Use energy-efficient products • Turn off appliances when not in use • Reduce heating by insulating buildings • Use double- or triple-glazed windows • Turn off taps when not in use • Walk more or ride a bicycle • Use less heating/air-conditioning

▲ Table 2.3.1. Methods of climate change mitigation

Government-led adaptation strategies for global climate change

Adaptation refers to initiatives and measures to reduce the vulnerability of human and natural systems to climate change.

There are many problems related to climate change, and many possible ways of adapting to them. Some of these are shown in table 2.3.2.

Climate-change risks	Potential adaptation strategies
• Flooding • Disease • Sea-level rise • Contaminated water • Dehydration • Drought • Famine/food shortages • Over-heating	• Early-warning systems • Emergency shelters • New forms of agriculture • Genetic engineering/high-yielding varieties of crops (HYVs) • Irrigation • Sea walls • Mosquito nets • Desalination • Migration

▲ Table 2.3.2. Risks of climate change and possible adaptation strategies

Civil society and corporate strategies to address climate change

WWF Australia

Australia is vulnerable to climate change. Most of its cities are coastal and low-lying. Extreme weather events have led to floods, droughts and fires. The Great Barrier Reef has suffered from coral bleaching. One in six species is threatened with extinction. Ocean acidification is threatening shellfish as well as coral. Water shortages and agricultural change are also each an increasing problem.

2.15 Briefly **explain** the terms "desalination" and "cost-benefit analysis". [2+2]

2.16 Outline one advantage and one disadvantage of desalination for Cape Town. [2+2]

2.17 Briefly **explain** why London should be better able to adapt to climate change than Cape Town. [2]

Test yourself

2.18 Define the terms (a) carbon-offset schemes, (b) geo-engineering, (c) carbon capture and (d) ocean fertilization. [1+1+1+1]

Test yourself

2.19 Identify the likely adaptation strategies that may be used for (a) sea-level rise, and (b) famine/food shortages. [2+2]

• **Civil society** – any organization or movement that works between the household, the private sector and the state to negotiate matters of public concern. Civil societies include non-governmental organizations (NGOs), community groups, trade unions, academic institutions and faith-based organizations.

>> **Assessment tip**

Try to keep up to date. Changes in government can have a significant impact on climate-change policy, for example. In 2017, the US president, Donald Trump, signed deals to increase production of fossil fuels, whereas the Chinese president, Xi Jinping, indicated that China would like the world to increase production and use of clean energy.

UNIT 2: GLOBAL CLIMATE—VULNERABILITY AND RESILIENCE

Content link

The success of civil societies in raising awareness of environmental risks is explored further in unit 6.3.

As a civil society, WWF Australia is committed to:

- limiting global warming to 1.5°C above pre-industrial levels by the end of the century
- achieving net-zero carbon pollution in Australia before 2050
- achieving 100% renewable energy in Australia before 2050, including 100% renewable electricity before 2035.

Test yourself

2.20 **Evaluate** the role of civil societies in the fight against climate change.

[5]

QUESTION PRACTICE

a) Study the following figure, which shows changes in atmospheric CO_2 and mean surface temperature since 1880.

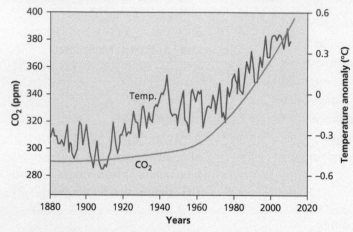

i) **Estimate** the increase in atmospheric CO_2 (ppm) between 1960 and 2015. [1]

ii) **Estimate** the change in temperature between 1960 and 2015. [1]

b) Draw a labelled diagram to show the main features of the greenhouse effect. [4]

c) i) **Suggest one** reason why predictions for global climate change vary. [2]

Essay

"Climate change impacts will be greatest for places with a high population density."
To what extent do you agree with **this** statement? [10]

How do I approach these questions?

a) i) A calculation is required—the 1960 value is approximately 300 ppm and the 2015 value is approximately 390 ppm. A value between 80 and 100 ppm would be accepted.

ii) This requires manipulation of the temperature anomalies—from between 0.0–0.1°C in 1960 to just under 0.6°C in 2015, so approximately 0.6°C (0.5–0.6°C accepted).

b) This requires a labelled diagram that explains how the greenhouse effect works.

c) Part (i) asks you to suggest reasons. You do not necessarily need to know the exact reasons, but you should be able to come up with some logical ideas, for example, about spatial scales, temporal scales, the role of feedback, the complexity of the issue.

Part (ii) asks you to outline reasons. You should give a brief explanation of how climate change can be caused by natural processes, for example, volcanic activity, dust storms, variations in solar output and so on.

Essay

As the command term is **to what extent**, your answer should provide supporting arguments and counterarguments for the statement given in the question. Essay questions in papers 2 and 3 are also synoptic, which means you will need information from across different units to answer the question. For example, you might want to draw on your knowledge of densely populated areas near coastlines (unit 1) to support the argument. However, you could counter the argument by describing how climate change impacts will also depend on latitude, and by discussing non-human impacts of climate change (unit 2).

a) i) 300 ppm to 395 ppm, so an increase of 95 ppm of CO_2

Mark 1/1

ii) 0.6°C

Mark 1/1

b)

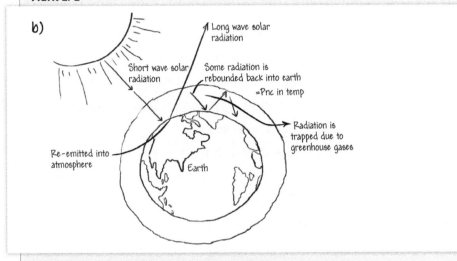

Long wave solar radiation

Short wave solar radiation

Some radiation is rebounded back into earth

=Pnc in temp

Radiation is trapped due to greenhouse gases

Re-emitted into atmosphere

Earth

Marks gained for references to "short-wave solar radiation"; "radiation trapped due to greenhouse gases"; "long-wave radiation"; and "re-emitted into atmosphere".

Marks 4/4

c) i) Certain things influence the global climate such as natural disasters, for example volcanoes. It is difficult to predict how much the climate will change until after the event has happened. For example the Iceland volcano reduced solar radiation into the Earth, making the climate cooler.

Another reason why predictions of climate change can vary is due to the fact that lots of things influence the climate of an area such as ocean circulation and air pressure, and these are constantly changing and so it is hard to predict future climate conditions accurately.

▼ More detail needed—data? By how much cooler?

▼ Ocean circulation and air pressure not developed enough to explain how they affect global climate change

One factor and poor choice of support material—it had very limited impact on global climate. The eruption of Mt Pinatubo would have been better as support.

Mark 1/2

ii) Solar flaring is a natural cause of climate change which involves an increase in radiation received from the Sun, therefore changing the climate. Sun spots are the opposite of solar flaring which is when dark patches appear on the surface of the Sun.

Only one natural cause explained. Sun spots not developed enough.

Mark 1/2

Essay

"Climate change impacts will be greatest for places with a high population density." **To what extent** do you agree with this statement?

▲ Valid point

▲ Good exemplification of a number of factors influencing the impacts of global climate change

▲ Summary point

▲ Valid point and example

▲ Good development of point

▲ Impacts of global climate change on a rich country

▲ Good development—changes from a national scale to global megacities

▲ Valid point—the ability to cope (afford) climate change adaptation schemes

Climate change is one of the most important environmental issues of our age. However, its impacts are not evenly spread around the world. Its impacts vary with population density, relief (height above sea-level), proximity to the oceans (coastal locations) and level of wealth (poverty – being able to plan for climate change) as well as demographic factors e.g. age, health, number of dependents. It is a complex issue.

Climate change could potentially have a huge impact on low-lying countries, such as Bangladesh. For example, a 1-metre rise in sea level would flood over 10% of the country and affect 9% of the population. Rising sea levels in the UK's coastal waters would affect London and the South East, and the Severn Estuary and the Mersey Estuary. These areas contain many large and important cities. Globally, there are many sites that are vulnerable to climate change e.g. New York (which was affected by Superstorm Sandy in 2012), Shanghai (which is built on land less than 5m above sea level, and Tokyo, where much of the city is built on land reclaimed from the sea. However, although these cities have the potential to be affected, they also have the resources to adapt to climate change. London has the Thames Barrier to protect it against tidal flooding and New York is raising its sea walls and building a new barrier to deal with storm surges.

Not everywhere is so lucky. Small islands e.g. Kiribati, are vulnerable to rising sea level but do not have the resources to construct sea defences of a sufficient size and strength. Populations living in slums low-lying cities are particularly vulnerable. In Jakarta, about 40% of the city is below sea level, and the poorest live close to river banks, canals and drainage areas, making them especially vulnerable to flooding. In addition to flooding, they are vulnerable to freshwater contamination (by seawater) and the spread of diseases such as malaria and dengue fever.

However, the impact of climate change varies spatially. The largest temperature rises since 2000 have been in the Arctic, at over 10°C. This has had a major impact on the ice cover, animal migrations, forest fires etc. but relatively little impact on humans, as the region generally has a low population density. Overall, it is not true to say that climate change will be greatest in areas of high population density. In some cases it may be so, especially in areas with a lack of resources to cope with the impacts of climate change, such as in the slums of Jakarta, but other impacts may be greater on the natural environment in areas of low population density, such as in the Arctic.

- ▲ Start of counterargument
- ▲ Valid point and support
- ▲ Valid point—vulnerability of slum populations
- ▲ Good detail
- ▲ Good detail—range of hazards
- ▲ Another counterargument
- ▲ Valid point
- ▲ Valid point— impacts on the natural environment
- ▲ Conclusion summarises argument and shows that it varies spatially (places) and due to wealth (power)

Good account—focused. Lots of valid support. Covers both sides of the argument, and has a conclusion that shows variations in place and power.

Marks 9/10

3 GLOBAL RESOURCE CONSUMPTION AND SECURITY

This unit examines how population growth and the expansion of the world's middle class have affected consumption of resources (for example, water, energy and food) to the point where there are issues of water, food and energy security in some regions. Nevertheless, there are possibilities to manage the world's resources sustainably, such as resource stewardship and the circular economy.

You should be able to show:

✔ how global development **processes** affect resource availability and consumption;

✔ how pressure on resources affects the future security of **places**;

✔ that there are **possibilities** for managing resources sustainably and **power** over the decision-making process.

3.1 GLOBAL TRENDS IN CONSUMPTION

- **Biocapacity** – the land and water to provide resources for humanity.

- **Ecological footprint** – the hypothetical area of land required by a society, a group or an individual to fulfill all of their resource needs and assimilate all of their waste. It is measured in global hectares (gha).

- **Embedded (virtual) water** – the amount of water used in the production and transport to market of goods.

- **Hydrocarbons** – chemical compounds consisting of carbon and hydrogen, such as oil and natural gas.

- **Green water** – the rainfall that is stored in the soil and evaporates from it; the main source of water for natural ecosystems, and for rainfed agriculture, which produces 60% of the world's food.

You should be able to show how global development processes affect resource availability and consumption:

✔ Global and regional/continental progress towards poverty reduction, including the growth of the "new global middle class";

✔ Measuring trends in resource consumption, including individual, national and global ecological footprints;

✔ An overview of global patterns and trends in the availability and consumption of:

  ✔ Water, including embedded water in food and manufactured goods;

  ✔ Land/food, including changing diets in middle-income countries;

  ✔ Energy, including the relative and changing importance of hydrocarbons, nuclear power, renewables and new sources of modern energy.

Global and regional/continental progress towards poverty reduction, including the growth of the "new global middle class"

One of the main successes of the millennium development goals (MDGs) was the global reduction in extreme poverty between 2000 and 2015. In 1990 around 50% of people in LICs lived on less than US$1.25 a day; by 2015 it was around 14%. In contrast, the number of people classified as middle class—that is, living on at least US$4 a day—almost tripled between 1900 and 2015. This population accounted for 18% of people in LICs in 1900 and nearly 50% in 2015.

The growth of the "new global middle class"

According to the World Bank (2016):

- Low-income economies are defined as those with a GNI per capita of US$1,045 or less in 2014.

- Middle-income economies are those with a GNI per capita of US$1,046–US$12,735.

- Lower-middle-income economies have an income of US$1,046–US$4,124.

- Upper middle-income countries have income of US$4,125–US$12,735.

- High-income economies are those with a GNI per capita of US$12,736 or more.

- Lower-middle-income and upper-middle-income economies are separated at a GNI per capita of US$4,125.

▼ Table 3.1.1. Size of the middle class, 2009–2030 (millions of people and global share)

	2009 (Millions)	(%)	2020 (Millions)	(%)	2030 (Millions)	(%)
North America	338	18	333	10	322	7
Europe	664	36	703	22	680	14
Central and South America	181	10	251	8	313	6
Asia-Pacific	525	28	1740	54	3228	66
Sub-Saharan Africa	32	2	57	2	107	2
Middle East and North Africa	105	6	165	5	234	5
World	1845	100	3249	100	4884	100

Source of data: Kharas, H., *The emerging middle class in developing countries*, World Bank (2011)

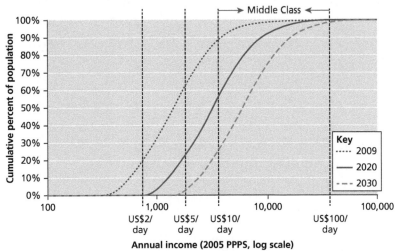

▲ Figure 3.1.1. Changes in annual income in China, 2009–2030

Measuring trends in resource consumption

As individuals and countries become wealthier, their consumption of resources increases. This includes food, water, energy and consumer goods, for example. Changes in diet, with increasing consumption of meat and dairy products, have an impact on the amount of water and energy used in agriculture. One way of examining resource consumption is to look at ecological footprints.

- **Blue water** – renewable surface water runoff and groundwater recharge; the main source for human withdrawals and the traditional focus of water resource management.
- **Grey water** – waste water that has been produced in homes and offices. It may come from sinks, showers, baths, dishwashers, washing machines, etc., but it does not contain fecal material.

Test yourself

3.1 Compare the relative distribution of middle-class population between 2009 and 2030. [3+3]

Test yourself

3.2 Identify the type of scale used in figure 3.1.1, and state why it has been used. [1+2]

3.3 Outline the changes in projected annual income in China, as indicated in figure 3.1.1. [2+2+2]

Concept link

PROCESSES: The rise of the new global middle class can be attributed to a number of geographical processes that have enabled people to escape from poverty and to join a segment of the global demographic. This is increasing the ecological footprint due to rising consumption.

Ecological footprints

The ecological footprint is the theoretical measurement of the amount of land and water that a population requires to produce the resources it consumes and to absorb its waste under prevailing technology. The ecological footprint tracks the use of six things: cropland, grazing land, fishing grounds, built-up land, forest area and carbon.

On the supply side, a city, state or nation's biocapacity represents the productivity of its ecological assets (including cropland, grazing land, forest land, fishing grounds and built-up land). These areas, especially if left unharvested, can also absorb much of the waste we generate, especially our carbon emissions.

Both the ecological footprint and biocapacity are expressed in global hectares—globally comparable, standardized hectares with world average productivity.

Rank	Country	EFP
1	United Arab Emirates	10.68
2	Qatar	10.51

▲ Table 3.1.2. Countries with the highest ecological footprint (EFP)

Rank	Country	EFP
187	Timor-Leste	0.49
188	Eritrea	0.48

▲ Table 3.1.3. Countries with the lowest ecological footprints

	Barcelona (4.52 gha)	Cairo (2.85 gha)
Food	33.3%	35.0%
Housing	4.7%	13.9%
Personal transportation	20.6%	12.2%
Goods	13.3%	9.4%
Services	3.9%	8.9%
Government	6.1%	7.2%
Infrastructure investment (houses, bridges, roads, factories)	18.1%	13.3%

▲ Table 3.1.4. Composition (%) of ecological footprints for Barcelona and Cairo

An overview of global patterns and trends in the availability and consumption of water, land/food and energy

Patterns and trends in the availability and consumption of water

Annual water availability is highest in Asia. Asia also has the highest annual consumption, at around 1,350 km³ per year. North America has the next highest water availability and consumption, followed by Europe. Both Africa and South America have much smaller consumption rates, while Oceania has the lowest availability and consumption rates.

Patterns and trends in the availability and consumption of food

Economic development is normally accompanied by improvements in a country's food supply. Increasing urbanization will also have consequences for the dietary patterns and lifestyles of individuals, not all of which will be positive. Changes in diets are referred to

Test yourself

3.4 Describe the main characteristics of the countries with the highest ecological footprints. [2]

3.5 Outline the main characteristics of the countries with the lowest ecological footprints. [2]

Test yourself

3.6 Study table 3.1.4. Using an appropriate data presentation technique, plot the urban ecological footprints for Barcelona and Cairo. **Compare** the main differences between the two footprints. [2+3]

3.7 Suggest why Barcelona has a higher biocapacity than Cairo. [2]

3.8 Briefly **explain** two reasons why the overall ecological footprints in urban areas are higher than in rural areas. [2+2]

3.9 Identify, and **justify**, one component of the ecological footprint that may be smaller in urban areas compared to rural areas. [1+2]

Content link

Factors affecting water availability are explored further in option A.3.

as the "nutrition transition". The pace of these changes seems to be accelerating, especially in the low-income and middle-income countries.

The dietary changes that characterize the "nutrition transition" include both quantitative and qualitative changes. The adverse dietary changes include shifts in the structure of the diet towards a higher energy-density diet with a greater role for fat and added sugars in foods, greater saturated-fat intake, reduced intakes of complex carbohydrates and dietary fibre, and reduced fruit and vegetable intakes.

Diets evolve over time, being influenced by many factors and complex interactions. Income, prices, individual preferences and beliefs, cultural traditions, as well as geographical, environmental, social and economic factors all interact in a complex manner to shape dietary consumption patterns.

The world has made significant progress in increasing food consumption per person. The growth in food consumption has been accompanied by significant structural changes and a shift in diet away from staples, such as roots and tubers, towards more livestock products and vegetable oils. However, this comes at a price—it requires far more water to produce meat and dairy products than it does to produce grain and vegetables. So water shortages are likely to become more frequent and intense as the demand for meat and dairy products increases.

▼ Table 3.1.5. Global and regional per capita food consumption (kcal per capita per day)

Region	Year 2015	Year 2030
World	2,940	3,050
Developing countries	2,850	2,980
Near East and North Africa	3,090	3,170
Sub-Saharan Africa[a]	2,360	2,540
Latin America and the Caribbean	2,980	3,140
East Asia	3,060	3,190
South Asia	2,700	2,900
Industrialized countries	3,440	3,500
Transition countries	3,060	3,180

[a]Excludes South Africa

Source of data: Food and Agricultural Organization of the United Nations (2002)

Patterns and trends in the availability and consumption of energy

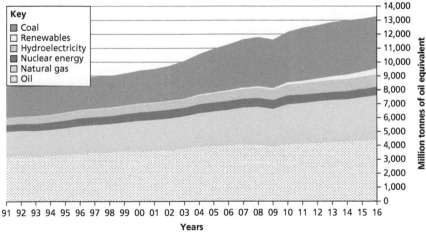

▲ Figure 3.1.2. World consumption of energy resources, 1991–2016

Source of data: BP Statistical Review of World Energy, June 2017

Test yourself

Study table 3.1.5.

3.10 Determine the region with (i) the greatest calorie intake and (ii) the lowest calorie intake in 2015.   [1+1]

3.11 Identify the regions predicted to experience the highest (i) absolute and (ii) relative increase in calorie intake between 2015 and 2030.   [1+1]

Test yourself

Look at figure 3.1.2.

3.12 Explain the scale "million tonnes of oil equivalent" (mtoe).   [2]

3.13 Identify the type of graph that is being used, and **explain** briefly why it is appropriate for this data.   [1+2]

3.14 Describe the main changes in the consumption of energy resources between 1991 and 2016.   [3]

Overall global oil production increased from around 65 million barrels per day in 1991 to over 90 million barrels per day by 2016. The largest relative increase was in the Middle East and South and Central America, but there were relative falls in Europe/Eurasia and North America, despite an absolute increase in the amount of oil produced.

Nuclear energy provides a relatively small amount of the world's energy. In 1991, it accounted for less than 500 mtoe out of a global total of approximately 8,000 mtoe (which is around 6%). By 2016, less than 5% of world energy consumption came from nuclear.

Nuclear energy peaked in 2006 but then fell around 2011 (possibly reflecting reaction to the Fukushima-Daichii nuclear disaster in Japan).

New sources of modern energy

Biofuels are a type of modern energy source. They are made from plants grown today, whereas fossil fuels are made from plants and animals that died millions of years ago. For decades, Brazil has turned sugarcane into ethanol, and some cars there can run on pure ethanol or as an additive to fossil fuels. In 2016, United Airlines announced a new initiative to integrate biofuel into its energy supply in the hope of reducing greenhouse gas emissions by 60%.

3.2 IMPACTS OF CHANGING TRENDS IN RESOURCE CONSUMPTION

- **Recover** – capture some value (for example in an energy-from-waste plant or as a nutrient, such as compost).

- **Recycle** – use in a different way (may involve "downcycling").

- **Reduce** – use less resource to meet the need (also called "minimization").

- **Remove** – eliminate demand altogether (also called "prevention" or "conservation").

- **Re-source** – change materials or sources (for example, using low-carbon electricity to meet demand).

- **Resource stewardship** – a concept that humans can use resources in a way that is sustainable.

- **Return** – put back in an environmentally benign way. This may require treatment or containment.

You should be able to show how pressure on resources affects the future security of places:

✔ The water–food–energy "nexus" and how its complex interactions affect:

  ✔ National water security, including access to safe water;

  ✔ National food security, including food availability;

  ✔ National energy security, including energy pathways and geopolitical issues;

✔ The implications of global climate change for the water–food–energy nexus;

✔ The disposal and recycling of consumer items, including international flows of waste.

The water–food–energy "nexus"

The water–food–energy nexus refers to the interrelationships between these three economic sectors. For example, the water sector influences the food sector and vice versa. Water infrastructure and use impact land use (such as irrigation), food production and fish stocks. Food production may impact on water quality (such as eutrophication, salinization). Water influences energy, for example, hydroelectric power (HEP), water for cooling/cleaning. In turn, energy developments may influence water by altering water temperature, quality and availability for other purposes. Food production requires energy for machinery, lighting, transport, processing, etc. In return, food production may produce energy sources, such as biofuels, and help to reduce the negative impacts of burning fossil fuels, for example, by acting as carbon sinks. By promoting one type of development over another, the water available for food developments is reduced. Similarly, if water is used for food (irrigation), the potential for energy developments is reduced. The nexus shows that developments in one sector may have unintentional impacts on the other sectors. The water–food–energy nexus is central to sustainable development. Demand for all three is increasing due to rising population, increasing wealth, rapid urbanization, changes in diet and economic growth.

National water security

Despite the recognition of water–food–energy nexus linkages, current approaches to water management in many areas treat the three sectors independently. That is the case in the Southern African Development Community (SADC), where a lack of cooperation and coordination has hampered developments in the water sector. Moreover, many of the river basins are transboundary (have parts of their river system in different countries) which makes coordination more difficult. Most of the dams in the region were built for a single purpose although some, such as the Kariba Dam in Zambia, are now being adapted for multiple purposes.

There is an uneven distribution of water in the SADC. Up to 75% of the region receives less than 650 mm of rain per year. Although there are plentiful supplies in DR Congo, the infrastructure to redistribute it around the region does not exist. A large proportion of SADC's population lives in rural areas (for example, 67% in Zimbabwe and 59% in Zambia), and access to safe water is limited. In Zimbabwe, around 25% of households have to pay for their water. Even in South Africa, the richest country in the region, around 5 million people lack access to clean water. In addition, climate models forecast a decrease of about 20% in annual rainfall by 2080 in southern Africa. With increasing demand for water from agriculture and energy, the availability of water for domestic consumption will be squeezed.

National food security

Within the SADC region, only around 6% of the land is cultivated, but this sustains the livelihood of 60% of the population. Only 3.5% of the region's arable land is irrigated, and this increases the vulnerability of the region to food insecurity. During the drought of 2015–16, over 40 million people became food insecure. Nevertheless, agriculture contributes some 17% of the regional GDP. Despite the importance of agriculture in the region and its consumption of 76% of the region's water resources, the current agricultural performance is insufficient to ensure regional food security and economic growth. Reasons for the low growth include low investment, regular droughts, lack of credit and poor farming practices. In addition, climate change is causing increased rainfall variability and reductions in crop yields. As more water is needed for energy developments or domestic/industrial uses, there may be increased pressure on the agricultural sector's supply of water. The area is likely to experience more food shortages, and reductions in food availability despite its growing population.

National energy security

The SADC faces energy insecurities. Some dams, such as Kariba, which were originally constructed for the purpose of hydroelectric

- **Reuse** – reintroduce into the same method as before.
- **Energy security** – having access to sufficient, clean, reliable and affordable energy sources for cooking, heating, lighting, communications and productive uses.
- **Food security** – having a sufficient amount of good-quality food.
- **Nexus** – the interrelationship, interdependence and interactions between water, food and energy.
- **Water security** – continuing access to safe drinking water and sanitation.

▲ Figure 3.2.1. Human pressure on water resources in the Eastern Cape, Republic of South Africa

Concept link

PLACES: Some places are more secure than others in terms of their access to food, energy and water. The resources could be sourced within the country's borders, or alternatively they can be secured from other places. The spatial interaction between countries and their ability to trade enables the latter.

production, how been diversified to provide for aquaculture, urban water supply, ecotourism, transport and mining activities. SADC has large energy resources that are relatively unexploited. There are 15 transboundary river basins which could be developed for HEP, although cooperation between countries is not guaranteed. Angola, DR Congo, Mozambique and Zambia have the capacity to supply the whole region with electricity, if not the funds. However, biomass remains the main source of energy, as only 24% of the total population and 5% of the rural population have access to electricity. Over-dependence on biomass has led to large-scale deforestation and desertification in the region.

Demand for energy is increasing due to urbanization, industrialization and economic growth, and that has led to frequent power blackouts in the region.

The pressure to produce more food and energy under increasing water scarcity requires careful management, coordination and cooperation among the three sectors. However, the predicted decline in rainfall totals will impact on energy production, food production and access to safe water.

Countries in the Middle East control about 50% of the world's remaining oil reserves. This gives the Middle East an economic and political advantage—countries that want oil may have to stay on friendly terms with those that supply it. Countries that depend on the region for their oil need to:

- help ensure political stability in the Middle East
- maintain good political links with the Middle East
- involve the Middle East in economic cooperation.

On the other hand, the situation is also an incentive for rich countries to increase energy conservation or develop alternative forms of energy.

Content link

Oil policies of Middle Eastern countries are discussed further in unit 4.1.

The implications of global climate change for the water–food–energy nexus

Climate change could influence the water–food–energy nexus in many contrasting ways. In some areas it may reduce agricultural productivity, whereas in other areas it may increase it. Water supplies will diminish in some areas and increase in others. The demand for energy will also change. Climate change is expected to increase the frequency of climate-related shocks, and these will have an impact on food, water and energy supplies. Moreover, due to their interconnections and interdependence, an impact on one part will have an influence on the other two.

Attempts to limit climate change may also have an impact on the water–food–energy nexus. The production of biofuels and hydroelectric power may create new demands for water resources. Some methods of adaptation to climate change, such as the use of drip irrigation and desalination of seawater, are very energy intensive. Increased groundwater use would also require extra pumping and therefore energy resources.

Global climate change creates critical challenges for water, energy and food, with increasing temperature, reducing snowpack and changing precipitation, as well as ecosystem processes at regional scales. In the Sacramento–San Joaquin Delta and Central Valley

Content link

The consequences of global climate change are discussed in unit 2.2.

watersheds in California, USA, the ecosystem services are reduced due to the increased regional temperature, changes in snowpack and precipitation, and increased water stresses from drought. The reduced services affect the water and energy nexus and agricultural food production, as well as fish and wildlife habitats.

Likely impacts include:

- Projected temperature increase ranging from 2 to 5°C by 2100;
- Loss of snowpack, with 48–65% of snow water content loss by the end of this century;
- Droughts, with more dry years and less water, which will affect food and energy;
- More frequent flooding and fire, affecting water quality in the watershed;
- Rising sea levels;
- Increasing energy demand;
- Changes in species and habitats.

Countries with contrasting levels of resource security

Two countries with very different levels of resource security are Saudi Arabia and Yemen. Saudi Arabia is currently water-, food- and energy-secure. It relies on its oil-based economy to import food and to cover the costs of desalination of seawater. Its water consumption is very high per person, and it has depleted its groundwater reserves.

By 2050, Saudi Arabia's population is forecast to be 50 million, up from 29 million in 2015. This will place great pressure on food and water production. Rapid urban growth and improved living standards have influenced the demand for food and water, and the country has experienced a nutrition transition.

Around 97% of Saudis have access to safe water. Average water consumption is 100–350 litres per day in urban areas, and 15–20 litres per day in rural areas. Despite its limited resources, Saudi Arabia produces and exports dates, dairy products, eggs and some fruit and vegetables. Agriculture has been heavily subsidized by the government in the past, although the levels of subsidies have been reduced.

In contrast, Yemen is the most food-insecure country in the Middle East and has one of the world's highest rates of hunger. It imports about 60% of its food. Over 10 million Yemenis, over 40% of the population, are food-insecure. Its population, 26 million, is expected to double by 2040.

Yemen's food and water crisis is linked to its political and social instability. However, scarce natural resources, including water, have increased this instability. Conflict has exacerbated food insecurity. Population displacement means that some farmers cannot harvest the food they have grown, and have no income to buy food at a market. Water scarcity reduces crop yields. The production of qat, a narcotic drug, is widespread and accounts for about 40% of Yemen's agricultural water consumption.

Yemen is one of the five most water-stressed countries in the world, with around 86 cubic metres of water available per person per year. Yemen's current food and water demand already exceeds supply. Moreover, due to the conflict between Saudi Arabia and Yemen, oil imports have been badly affected.

Test yourself

3.15 Analyse how, and why, global climate change may affect the water–food–energy nexus in California.   [2+2]

3.16 Outline two advantages and two disadvantages of global climate change for the water–food–energy nexus.   [2+2]

The disposal and recycling of consumer items

There are many forms of disposal and recycling of consumer items.

- Remove—the elimination of demand altogether

- Reduce the amount of waste

- Reuse goods to extend their lifespan, for example reuse of milk bottles or reuse of old tyres to reduce soil erosion

- Recycle—create new forms of the same product (recycled paper), or put used goods to another use (for example plastic bags used as bin liners; old clothes used as cleaning cloths)

- Recover value—compost biodegradable waste for use as fertilizer and/or incinerate (burn) waste; collect electricity and heat from it

- Disposal—put waste in landfill sites (natural or the result of quarrying) or use to make artificial hills.

▲ Figure 3.2.2. Recycling is becoming more widespread in many HICs

Flows of waste

Increasing amounts of waste are being exported internationally. In general, the flow of waste (including electronic waste) is from HICs to LICs and MICs. The European Environment Agency estimates that between 250,000 tonnes and 1.3 million tonnes of used electrical products are shipped out of the EU every year, mostly to west Africa and Asia. Research by the Massachusetts Institute of Technology (MIT) suggests that, in 2010, the US discarded 258.2 million computers, monitors, TVs and mobile phones, of which only 66% were recycled.

Test yourself

3.17 Contrast two positive and **two** negative impacts of the disposal of e-waste. **[2+2]**

3.3   RESOURCE STEWARDSHIP

Concept link 🔗

POWER AND POSSIBILITIES: Resource stewardship is a pathway to achieving environmental sustainability and involves a range of stakeholders. This complex web of vested parties involves multi-governmental organisations such as the UN with their SDGs and national governments altering their mindsets and policies to engage with the circular economy. People also have everyday choices to make regarding resources. As environmental degradation, consumerism and population sizes all increase, the decisions made at a number of levels and scales will determine the future of the planet and its citizens.

You should be able to show examples of possibilities for managing resources sustainably and power over the decision-making process:

✔ Divergent thinking about population and resource consumption trends:

    ✔ Pessimistic views, including neo-Malthusian views;

    ✔ Optimistic views, including Boserup;

    ✔ Balanced views, including resource stewardship;

✔ Resource stewardship strategies, including:

    ✔ The value of the circular economy as a systems approach for effective cycling of materials and energy;

    ✔ The role of the UN Sustainable Development Goals and progress made toward meeting them.

Divergent thinking about population and resource consumption trends

There are many views on the relationship between population and resources. Two of the most famous are the views of Thomas Malthus and Ester Boserup.

Malthus wrote in 1798 but his main idea—that population growth would outstrip the growth of resources—has been updated by neo-Malthusians, such as Paul Ehrlich and the Limits to Growth team. The neo-Malthusians have the same pessimistic message, but they encourage the use of contraception and family planning as a way of reducing population growth. Malthus himself was against the use of artificial contraception, as he considered it to be immoral (he was a vicar). Neo-Malthusians also suggest that there needs to be greater redistribution of wealth and fairer access to resources.

In contrast, Ester Boserup suggests that, as the need arises, people will find solutions to overcome shortages of resources. She suggested that new techniques/methods would increase productivity, for example, using the land more intensively, increasing the use of irrigation and fertilizers, and using high-yielding varieties of crops.

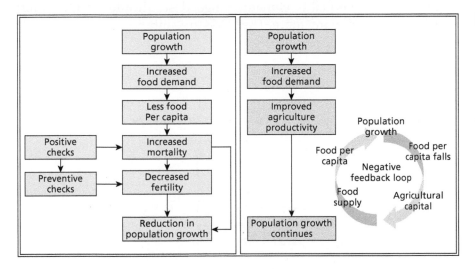

▲ **Figure 3.3.1.** Malthusian (left) and Boserup (right) views on population growth

The neo-Malthusians have been predicting global famine because of an overpopulated planet. In *The Population Bomb*, Ehrlich pronounced: "The battle to feed all humanity is over" (Paul Ehrlich, *The Population Bomb*, 1968, Buccaneer Books). Neo-Malthusians predicted a crisis of food, energy, water and land resources.

According to the economist Amaryta Sen, famine was not caused by declining availability of food, but a decline in food entitlement, as a result of unemployment, a lack of benefits, rising food prices, etc. For example, many landless labourers in Kalahandi, India, may help to produce rice but cannot afford to buy it. However, Stephen Devereux of the International Development Institute believes that it is more than an economic factor for example, political regime, natural disasters, civil war, local and global governance, climate change and environmental issues are all important in different places.

The term "stewardship" refers to the careful management of the environment and its resources, in a way that benefits humanity and is sustainable. "Stewardship" suggests that people are caretakers—looking after the environment and its resources for the benefit of

---

- **Divergent thinking** – finding new solutions to old problems; thinking "outside the box".

- **Circular economy** – an economy that preserves natural capacity, optimizes resource use and reduces loss through managing finite stocks and renewable flows.

- **Refurbish** – to restore and to make useful again.

- **Recycle** – refers to the manufacturing of a used good into another good that can be used again, such as bottles, paper, aluminium. However, not all products can be recycled. Coffee cups made from cardboard with plastic coating cannot be recycled.

- **Biochemical feedstock** – refers to the production of renewable energy from crops such as corn, sugarcane, soyabeans and palm oil.

- **Anaerobic digestion/ decomposition** – the production of biogas and/or fertilizer from crops.

Test yourself

3.18 Distinguish between the Malthusian and neo-Malthusian views of population growth and resources. [2+2]

≫ Assessment tip

Make sure that you answer the question! If a question asks about neo-Malthusian solutions to the population–resources issue, make sure that you provide neo-Malthusian solutions rather than Malthusian ones.

Test yourself

3.19 Briefly **describe** how increased demand for food could lead to improvements in agricultural productivity. [3]

3.20 Outline the contribution of academics, such as Sen and Devereux, to the debate about population and resources. [2+2]

3.21 Explain the term "resource stewardship". [2]

3.22 Suggest how resource stewardship contributes to the management of population and resources. [2]

3.23 Define "circular economy". [1]

3.24 Suggest how materials from animals used for the production of meat and milk could be used for biochemical feedstock. [2]

humanity, rather than just preserving environments for their own benefit. It is a concept that operates on a global scale, and, as such, is very difficult to achieve. Examples of global resource stewardship include attempts in reaching agreements on things like climate change, marine fishing policies and reducing plastic pollution.

Resource stewardship strategies

The circular economy

A circular economy is one that preserves natural capacity, optimizes resource use and reduces loss through managing finite stocks and renewable flows. It is an economy that restores and regenerates resources, and keeps products, materials and components at their highest utility and value.

The role of the UN Sustainable Development Goals (SDGs)

The SDGs were introduced in 2015 and will run until 2030. They follow on from, and extend the original, Millennium Development Goals that existed between 2000 and 2015. It is too early to assess how the SDGs are faring. There is still much poverty around the world, and problems related to gender equality, climate change, desertification, acidification of the oceans and many more. The SDGs are ambitious.

QUESTION PRACTICE

The following graph shows the size and composition of the ecological footprint for selected countries.

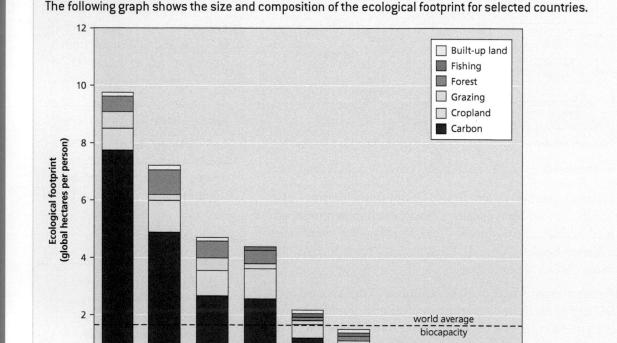

a] **Define** the term "ecological footprint". [2]

b] **Outline two** differences in the ecological footprint of the USA and Nigeria. [2 + 2]

c] Briefly **explain** two factors that lead to a decline in food availability. [2 + 2]

Essay

"By the end of the 21st century there will be too many people and not enough resources left."
To what extent do you agree with this statement? [10]

How do I approach these questions?

a) This asks for a definition. It is worth 2 marks, so you should have two specific points in your answer.

b) You are required to give two differences in the ecological footprint of the USA and Nigeria. One difference could be the size of the ecological footprint and another could be the composition. You will need to add details.

c) An explanation is required. You should identify two valid factors and then develop each one further and/or give some exemplification.

a) The theoretical measurement of the amount of land and water that a population needs to produce the resources it consumes and to absorb its waste measured in global hectares.

▲ Good point

▲ Valid point

A good definition.
Marks 2/2

b) The USA produces much more carbon than Nigeria does, 5 ghas per annum in the USA compared with just above zero in Nigeria. This is because the USA uses more energy and therefore more fossil fuel as it is a more economically developed country and it has a higher population. The use of more fossil fuels emits more carbon, and uses more carbon. The USA also has a higher proportion of forest use of about 1 gha. This is because forest products are used for fuel, infrastructure and furniture. The USA needs more of this due to its high population and standard of living. Overall, the USA's ecological footprint is bigger than Nigeria's at 7 gha for the USA and just over 1 gha/person in Nigeria, which is below the world average biocapacity.

▼ Inaccurate use of data

▼ Not valid as the EFP is given as gha per person

▲ Valid point

▲ Valid point

▲ 1 mark

Somewhat contradictory in places. Could have developed the composition of Nigeria's EFP. Some valid points are made.
Marks 3/4

c) Physical factors and human factors cause food shortages. One of the physical factors that cause food shortages is the climate. If there is a drought, that is 29 consecutive days without rainfall, then the crops do not get enough water to grow. This is a very common problem in Sub-Saharan Africa.

▲ Definition of drought

▲ Development

Physical factor correctly explained, and some development, e.g. location, and definition of drought.

▲ Correct identification of human factor

▲ Located example

A human factor that can cause food shortages is civil war. For example, in Yemen they import 90% of their food. But because of the civil war, there was a blockade which prevented weapons being sent over but also food. This caused famine throughout their country as there was hardly any food available.

Human factor correctly identified/explained and located example – very contemporary

Marks 4/4

Essay

"By the end of the 21st century there will be too many people and not enough resources left." **To what extent** do you agree with this statement?

▲ Starts with a statistic

▲ Links to relative growth

▲ Increase in quality of life as well as quantity of people

▲ Suggests a population 'dilemma'

▲ Identifies the main problem

▲ Clear range of issues regarding resource use

▲ Very clear introduction

▲ Good to cover the nexus

▲ Exemplification of concepts

By the end of the 21st century, the world's population is projected to be around 11 billion, up from 7 billion at present. This is an increase of over 50%. More importantly, the world's population will have more middle income countries (MICs) than at present and fewer low income countries (LICs). If the Sustainable Development Goals (SDGs) are to be achieved, poverty will be eradicated by 2030 and everyone will be living in MICs or HICs. This may be good news for the world's population but it may be bad news for resource depletion. As people become richer they consume more resources. Richer people eat more meat and dairy products than poorer people; they use more water (showers, baths, water the garden, wash the car) and consume more energy (more electrical goods, private vehicles). More land, water and energy resources will be needed to fuel the increased demand for consumer and non-consumer goods.

The water-food-energy nexus is the concept that explains how increased demand and use of one component of these three resources has an impact on the other two. For example, if more energy is needed to produce food, there is less energy available for water production (e.g. desalination, pumping of groundwater, HEP). This shows that there could be a problem with resource security in the future.

The neo-Malthusians, such as Paul Ehrlich and the Club of Rome, suggest that population growth has the potential to outstrip the growth of resources, and that ultimately there will be a population crash. However, with population control, an increase in food supply, and redistribution of wealth, this could be avoided. However, reduced population growth is usually associated with higher resource consumption, due to greater wealth.

In contrast, Ester Boserup believed that people have the resources (knowledge and technology) to find a solution to the problems. She believed that increasing population size stimulated changes in agricultural techniques. At present, this could be GMOs, in-vitro meat production (stem cells) and greater use of hydroponics.

Other theories include Hardin's 'Tragedy of the Commons' i.e. common resources such as the oceans will be over-fished through greed and a lack of proper management. Resource stewardship suggests that resources can be used in a way that can be made available to future generations. It suggests environmental sustainability and social equity. The idea of a circular economy maximises resource use, preserves natural capital and reduces loss. It is possible that with more re-use and recycling of goods, waste disposal can be reduced and the life of natural resources extended.

Thus it is not inevitable that there will be too many people or not enough resources. However, it will take a major shift in the way we use resources – and waste them – if the world is to reach 2100 and have a high standard of living for all.

▲ Considers neo-Malthusian view

▲ Solutions offered

▲ Counter-view

▲ Range of new technologies

▲ Goes beyond normal Malthusian vs Boserup argument

▲ Introduces sustainability and equity

▲ Good range of concepts and exemplification

Conclusion brings it all together—touches on 'inevitability' but shows that the population-resource balance could go either way. Perhaps a more in-depth account of one or more resources and their use, decline and/or alternatives could allow for greater evaluative comments.

Marks 9/10

4 POWER, PLACES AND NETWORKS (HL ONLY)

The study of global interactions has a broader perspective than a study of globalization. "Globalization" often focuses on the domination of Western culture on the world, whereas "global interactions" suggests a two-way and complex process whereby cultural traits and commodities may be adopted, adapted or resisted by societies. The process is neither inevitable nor universal.

You should be able to show:

✔ how global **power** and influence varies spatially;

✔ how different **places** become interconnected by global interactions;

✔ how political, technological and physical **processes** influence global interactions.

4.1 GLOBAL INTERACTIONS AND GLOBAL POWER

• **Globalization** – "the growing interdependence of countries worldwide through the increasing volume and variety of cross-border transactions in goods and services and of international capital flows, and through the more rapid and widespread diffusion of technology" (International Monetary Fund).

• **Soft power** – this refers to the positive influence that one country can have over another through its culture, education, enterprise, digital expertise, engagement and government, as opposed to **hard power** where force or coercion is used.

• **Superpower** – a nation or group of nations that has a leading position in international politics.

You should be able to show how global power and influence varies spatially:

✔ Globalization indices showing how countries participate in global interactions;

✔ Global superpowers and their economic, geopolitical and cultural influence;

✔ Powerful organizations and global groups:

    ✔ G7/8, G20 and Organization for Economic Cooperation and Development (OECD) groups;

    ✔ Organization of the Petroleum Exporting Countries' (OPEC) influence over energy policies;

    ✔ global lending institutions, including the International Monetary Fund (IMF) and New Development Bank (NDB).

Globalization indices showing how countries participate in global interactions

The KOF Index of Globalization

The KOF Index covers the economic, social and political dimensions of globalization:

• economic globalization—long-distance flows of goods, capital and services, as well as information and perceptions that accompany market exchanges (36% of the Index)

• social globalization—the spread of ideas, information, images and people (38% of the Index)

• political globalization—the diffusion of government policies (26% of the Index).

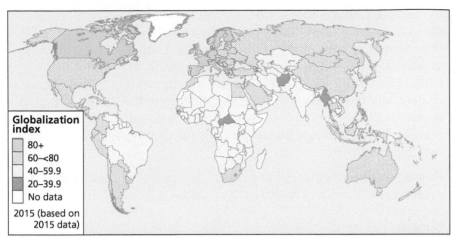

▲ Figure 4.1.1. The KOF Index, 2018 (based on data for 2015)

The level of globalization fell slightly in 2015. Overall, the Netherlands, Switzerland and Sweden were the most globalized countries in 2015, which is to say they had the most economic, social and political links to other countries.

From 1990 to 2007, the level of globalization increased rapidly. In 2015, the level of globalization decreased for the first time since 1975. This was caused by declining economic globalization and stagnating social globalization. The isolationalist policies of the USA in recent years and the outcome of the UK's Brexit referendum in 2016 have resulted in reduced economic globalization.

The KOF Index now distinguishes between de facto (actual/real) and de jure (legal) globalization. The potential for de jure globalization has increased, but de facto globalization has fallen—the cross-border exchange of goods and services has declined. De facto trade relates to trade in goods and services; de jure trade relates to customs duties, taxes and restrictions on trade. De facto social globalization refers to international telephone connections, numbers of tourists and migration, whereas de jure social globalization refers to freedom of the press and international internet connections. De facto political globalization refers to the number of embassies and international non-government organizations (NGOs), whereas de jure political globalization focuses on membership of international organizations and international treaties.

Due to their lesser degree of interdependence, larger countries tend to be placed much lower down the rankings in the KOF Index than smaller countries. For example, the USA was in 63rd position for economic globalization, 29th for social and 10th for political. Due to their high level of economic, social and political interdependence within the EU, the largest European economies (Germany, Italy, France and the UK) are overall considered more globalized than the USA or China.

The most economically globalized countries tend to be those that operate as financial hubs and/or trading centres, for example, Singapore, the Netherlands and Malta.

Global superpowers and their economic, geopolitical and cultural influence

The USA remains the world's main superpower. It has a large economy, strong allies and partners, and a massive military superiority. It has considerable "soft power" through its film and TV industries, and universities. Its economy has been helped since 2000 by the discovery of huge amounts of shale gas.

>> **Assessment tip**

When referring to a map such as figure 4.1.1, make sure that you refer to the information in the key. For example, the USA and Australia both have a KOF Index of 60–<80.

Test yourself

4.1 Define globalization. [2]

4.2 Explain the advantages of using an index of globalization, such as the KOF Index. [6]

4.3 Explain the disadvantages of using an index of globalization, such as the KOF Index. [6]

>> **Assessment tip**

You need to be able to discuss detailed examples of at least two contrasting superpowers.

In contrast, China is the world's second largest superpower and is the greatest long-term challenger to the USA. The government of US President Trump rejected the concept of the Trans-Pacific partnership in 2018, and refused to act on climate change. In 2017, however, China announced that it may lead the way in tackling global climate change, and has been expanding its influence in the Pacific Rim both economically and militarily. The Chinese President, Xi Jinping, has also managed to consolidate his power over the country, and has a vision for Chinese growth.

Russia has also re-emerged as a significant power. It participated in the protection of President Assad in Syria and it annexed Crimea.

Hard and soft power

Hard power refers to the power that countries may exert through force or coercion. A country's military strength is a good indication of its hard power, but so too is its economic power, and the ability to establish trading deals in which it benefits.

Soft power refers to the positive influence that one country can have over another through its culture, education, enterprise, digital expertise, engagement and government. Cultural soft power can be spread through the film industry, music, museums and galleries. Educational soft power is generally measured in terms of higher education, through the quality of the university, its ability to attract international students and its contribution to academic research publishing. Enterprise refers to the attractiveness of a country's business model, its capacity for innovation and its tax framework. The digital component of soft power relates to the influence of the world's leading high-tech companies. Engagement refers to the best-networked states, membership of multilateral organizations and embassy networks. The role of government in soft power relates to freedom, human rights, democracy and equality.

Powerful organizations and global groups

G8/G7

The G8 is a group of eight leading industrial countries, consisting of Canada, France, Germany, Italy, Japan, Russia, the UK and the USA. They meet periodically to discuss major world issues such as world recession. However, following the Russian annexation of Crimea from the Ukraine, Russia was suspended and the G8 became the G7. Without Russia, the G7 countries have more in common with each other; but, without China, the G7 (or G8) countries cannot claim to be the world's leading economic and political powerhouses.

G20

The G20 is a grouping of 20 of the world's major economies that discuss policies related to financial stability. It includes representatives from Argentina, Australia, Brazil, Canada, China, France, Germany, India, Indonesia, Italy, Japan, Mexico, Republic of Korea, Russia, Saudi Arabia, South Africa, Turkey, UK, USA and the EU. Each year the G20 invites several guest countries to participate in G20 events and contribute to the agenda.

▲ **Figure 4.1.2.** The image of the Statue of Liberty has had significant cultural impact globally. This is an example of the soft power of the USA

Test yourself

4.4 Distinguish between "hard power" and "soft power". [2]

 Content link

Cultural influences across countries are examined further in unit 5.2.

The G20 countries account for 90% of the global economy, 80% of global trade, 66% of world population and 84% of fossil fuel emissions.

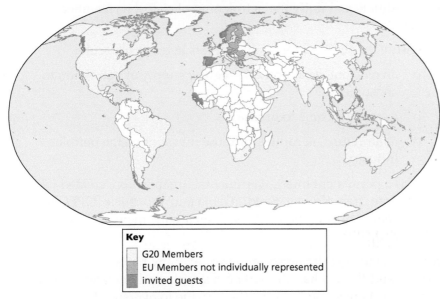

▲ Figure 4.1.3. The G20 members map

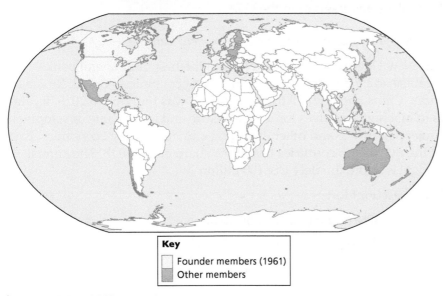

▲ Figure 4.1.4. OECD countries

The Organisation for Economic Co-operation and Development (OECD) was formed in 1961, and by 2017 there were 35 members. The G7 is its inner group of leading industrial nations. The OECD aims to identify, analyse and discuss problems, and to find potential solutions for them. Its aims include:

- to restore confidence in markets

- to foster and support new sources of growth through innovation and environmentally friendly "green growth"

- to develop skills for people of all ages to work productively.

Test yourself

4.5 Outline one advantage and **two** disadvantages of the G7 countries.   [1+2]

4.6 Distinguish between membership of the G7 and G20 countries.   [2]

4.7 Suggest the likely characteristics of the member countries of the OECD (Figure 4.1.4).   [2]

OPEC (Organization of the Petroleum Exporting Countries)

The role of OPEC is to:

- coordinate and unify the petroleum policies of its member countries

- ensure the stabilization of oil markets in order to secure:

    o   an efficient, economic and regular supply of petroleum to consumers

    o   a steady income to producers

    o   a fair return on capital for those investing in the petroleum industry.

OPEC is a permanent intergovernmental organization, created in 1960 by Iran, Iraq, Kuwait, Saudi Arabia and Venezuela. OPEC rose to prominence during the 1970s, with its involvement in the oil crises of 1973 and 1979.

In recent times, escalating social unrest in many parts of North Africa and the Middle East has affected both supply and demand of oil. Prices were stable between 2011 and 2014, but due to oversupply, prices fell in 2014. OPEC's influence has been further reduced by the discovery of huge shale gas reserves in the USA.

Global lending institutions: New Development Bank

The New Development Bank (NDB) is the first multilateral bank established by developing and emerging economies—Brazil, Russia, India, China and South Africa (BRICS). It was founded in 2015. Its aim is to mobilize resources for infrastructure and sustainable development projects in BRICS and other emerging and developing economies. Each of the five BRICS countries has a 20% share in the NDB. The annual budget is approximately US$100 billion.

BRICS account for:

- about 42% of the world's population

- over 20% of global GDP

- 27% of the world's land area.

Projects include:

- US$81 million for 100,000 kWh rooftop solar photovoltaic power project proposed to be implemented in the Shanghai Lingang Industrial Area (SLIA).

- US$300 million for an offshore wind power project proposed to be implemented in Fujian Province, China.

International Monetary Fund (IMF)

The IMF was established in 1944 with the aim of:

- promoting international monetary cooperation

- facilitating the expansion and balanced growth of international trade

- promoting international stability

- making resources available to members experiencing balance of payments difficulties.

Concept link

POWER: The influence that one country can have over another country changes over time and space. Some countries have more influence than others, and as such, are classified as global superpowers. In addition to these powerful countries, multi-governmental organizations, such as OPEC, have global superpowers and their members can create and resolve complex situations, such as the production and availability of oil, or the lending of capital to help alleviate the effects from an economic recession.

>> Assessment tip

When writing about multi-government organizations (MGOs), try to have some balance, that is, have some positive aspects as well as some negative ones.

The IMF's fundamental mission is to ensure the stability of the international monetary system. It does this by keeping track of the global economy and the economies of member countries, lending to countries and giving practical help. Unlike development banks, the IMF focuses on lending for debt reduction and export-led development.

The IMF has 189 members, and its biggest borrowers include Portugal, Greece, Ukraine and Pakistan. Members of the IMF pay a quota, based on their GDP. The quota that a country pays determines the maximum amount of finance a member can receive from the IMF and its voting power within the IMF.

> **Test yourself**
>
> **4.8 Suggest** definitions for the terms **(a)** bilateral surveillance and **(b)** multilateral surveillance. [2+2]
>
> **4.9 Distinguish** between "capacity development" and "lending". [2]
>
> **4.10 Evaluate** the role of the NDB. [4]

4.2   GLOBAL NETWORKS AND FLOWS

You should be able to show how different places become interconnected by global interactions:

✔ An overview of contemporary global networks and flows:

  ✔ global trade in materials, manufactured goods and services;

  ✔ an overview of international aid, loans and debt relief;

  ✔ international remittances from economic migrants;

  ✔ illegal flows, such as trafficked people, counterfeit goods and narcotics;

✔ Foreign Direct Investment (FDI) and outsourcing by transnational corporations (TNCs), and ways in which this networks places and markets;

  ✔ Two contrasting detailed examples of TNCs and their global strategies and supply chains.

An overview of contemporary global networks and flows

Global trade in materials, manufactured goods and services

In 1990, about 50% of world trade by volume was between HICs. A further 15–20% was from HICs to LICs, and up to 40% was from LICs to HICs. The volume of trade between LICs was relatively small.

By 1995, the main changes that had occurred were the decline in the relative importance of trade between HICs; a doubling of trade between LICs, and a relative decline in trade from HICs to LICs.

There was also an increasing influence of China as a global economic power. Since 1995, the proportion of trade between HICs has gone down to about 30%, trade between LICs has risen to around 20%, and trade between LICs and China has risen to around 10%. The volume of trade from HICs to China has also increased.

In terms of trade by value, in 1990, trade between HICs accounted for about 70% of world trade, trade between HICs and China a further 10%, and from LICs to HICs also about 10%. By 2015, the proportion of world trade accounted for by trade between HICs had fallen to around 45%, trade between HICs and China and from LICs to HICs still accounted for about 20% each, but there was an increase in the value of trade between LICs.

> • **Remittances** – the money sent back by a migrant to their family in the migrant's country of origin.
>
> • **Narcotics** – illegal drugs that are prohibited from general use.
>
> • **Trafficked people** – people who are moved against their will; for example, for forced labour or sexual slavery.
>
> • **FDI (Foreign Direct Investment)** – the investment by a company into the structures, equipment or organizations of a foreign country. It does not include investment in shares of companies of other countries.
>
> • **TNC (transnational corporation)** – a company with facilities in more than one country. Generally, decision-making and research and development take place in HICs, whereas assembly and production is more likely to occur in regions of low labour costs.
>
> • **Resource endowment** – the amount of land and physical resources (water, fertile soils, minerals, e.g. oil, natural gas, coal, iron ore) that a country has within its borders.

There are four main components of international trade in services:

- Cross-border trade—the delivery of a service from one country to another, for example, airlines flying between countries or the provision of accountancy services by another country

- Consumption abroad—for example, tourism, study abroad, overseas medical services

- Commercial presence—services provided by a supplier in another country, for example, banks operating in other countries

- Presence of individual people—people working overseas, such as doctors, oil engineers, architects.

In 2014, global trade in services totalled around US$4.8 trillion. Commercial presence accounted for about 55% of the total, cross-border trade 30%, consumption abroad 10% and presence of natural persons less than 5%. Services account for about 75% of economic output in the EU and 70% in the USA. The EU28 was the world's largest exporter and importer of services, with a trade surplus of €162.9 billion.

International aid

The main donors of development aid are rich countries in North America and Europe, Australia, New Zealand and Japan, while the main recipients are in poor countries. The highest levels of aid would appear to go to much of sub-Saharan Africa, Eastern Europe and Russia, and South-East Asia. The largest donors are the USA and Japan, although each donates less than 0.25% of their GNI. France and the UK are the next largest donors, donating less than 0.5% of their GNI. The largest donors, in 2017, in relation to GNI were Saudi Arabia (1.8%) and the UAE (1.26%).

International loans

The International Bank for Reconstruction and Development (IBRD) is the world's largest development bank and hopes to eliminate poverty by 2030. The International Development Association (IDA) is the largest multilateral source of concessional finance (lending that offers lower interest rates and/or longer repayment periods) to boost growth and cut poverty. The IDA lends only to nations with a very low per capita income. For such countries, loans are interest-free and allow long repayment periods. While the IBRD provides loans and assistance to mainly middle-income countries, the IDA helps the world's low-income countries.

>> **Assessment tip**

Be sure to specify whether you are writing about the absolute amount of aid that a country donates (usually in US dollars) or the relative amount (as a percentage of GNI). For example, the USA gives the largest amount in absolute terms, but not in relative terms.

Test yourself

4.11 Describe the main changes in global trade in manufactured goods since 1990. [4]

4.12 Identify the main characteristics of the countries that were **(a)** the top ten World Bank borrowers between 1945–2015, **(b)** the top ten IBRD borrowers in 2015 and **(c)** the top ten IDA borrowers in 2015. [2+2+2]

▼ Table 4.2.1 Top ten World Bank borrowers, 1945–2015 (US$ billion)

India	102.1
Brazil	58.8
China	55.8
Mexico	54.0
Indonesia	50.5
Turkey	38.0
Argentina	30.8
Pakistan	27.7
Bangladesh	23.5
Colombia	21.7

Source of data: World Bank

▼ Table 4.2.2 Top ten IBRD borrowers, 2015 (US$ billion)

India	2.1
China	1.8
Colombia	1.4
Egypt	1.4
Ukraine	1.3
Argentina	1.3
Turkey	1.1
Morocco	1.1
Indonesia	1.0
Poland	1.0

▼ Table 4.2.3 Top ten IDA borrowers, 2015 (US$ billion)

Bangladesh	1.9
India	1.7
Ethiopia	1.4
Pakistan	1.3
Kenya	1.3
Nigeria	1.0
Tanzania	0.9
Vietnam	0.8
Myanmar	0.7
Ghana	0.7

Debt relief

The Heavily Indebted Poor Countries (HIPC) and the Multilateral Debt Relief Initiative (MDRI) are the two main approaches to reducing debt in poor countries. Some 36 countries eligible for debt relief have graduated from the programme (Figure 4.2.1). To qualify for debt relief, countries must undertake economic and social reform to try and reduce poverty. The MDRI also helps countries make progress towards achieving the sustainable development goals (SDGs).

Content link

The SDGs are outlined in unit 5.1: Development opportunities.

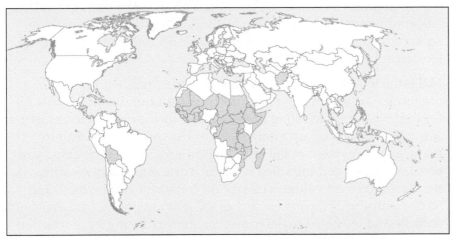

▲ **Figure 4.2.1.** Heavily Indebted Poor Countries (HIPCs), shaded in green

Test yourself

4.13 Describe the distribution of HIPC countries, as shown on figure 4.2.1. [3]

4.14 Explain the advantages of debt relief. [2+2]

International remittances from economic migrants

Remittances increased from almost zero in 1990 to around US$100 billion in 2002, and then increased rapidly to about US$450 billion in 2017. Although there was a dip in 2008–09, following the financial crisis/recession, the recovery was quick, and remittances continued to increase. Since 1998, remittances have been a more important transfer of money to LICs and MICs than overseas development aid.

Assessment tip

Remember that not all remittances flow from HICs to LICs and MICs—some are transferred from oil-rich countries to HICs and some from MICs to HICs.

Test yourself

4.15 (a) Suggest reasons why India, China, the Philippines and Mexico were the top remittance receivers of 2017. [2+2+2]

(b) In some countries, remittances account for a very high proportion of GDP. These include the Kyrgyz Republic, Tonga, Tajikistan, Haiti, Honduras and Moldova. **Outline** reasons why remittances should form such a high proportion of GDP in these countries. [2+2+2]

Illegal flows

There are many forms of illegal flows, such as trafficked people and narcotics.

Trafficking of people is a crime on a global scale. The UN Office on Drugs and Crime 2016 Report stated that between 2012 and 2014 there were over 63,000 victims of trafficking, from data provided by 106 countries. Trafficking can be domestic or international. About 60% of victims are international, although less than 30% are inter-regional. Most trafficking occurs within the same geographical sub-region. Frequently, victims come from relatively poorer countries and are exploited in relatively richer countries. The Middle East has the highest share of inbound trafficked people from other regions.

Drug trafficking is a major global trade involving the cultivation, manufacture, distribution and sale of illegal substances. The global drug trade is estimated to be worth more than US$300 billion, or 1% of total global trade.

Foreign Direct Investment (FDI) and outsourcing by transnational corporations (TNCs)

A transnational corporation (TNC) is an organization that operates in a large number of countries. Generally, TNC headquarters are in HIC cities, with research and development (R&D) and decision-making concentrated in growth areas of HICs, and assembly and production located in LICs and MICs.

FDI is the investment by a TNC into the structures, equipment or organizations of a foreign country. There are a number of benefits for the investor of investing abroad. These include access to cheaper raw materials and cheap foreign labour. However, it may lead to rising unemployment in the investing country, and an increased gap between skilled and unskilled workers. For the recipient country, there are increased volumes of exports, increased employment and foreign earnings. However, many of the jobs are poorly paid, have little security and may lead to the foreign country reducing its production of agricultural products as workers migrate to urban areas in search of the perceived well-paying jobs.

The Tata Group

The Tata Group comprises over 100 companies, encompassing cars and consulting, software and steel, tea and coffee, transport and power, chemicals and hotels. Tata Consultancy Services (TCS) is Asia's largest software company. Tata Steel is India's largest steelmaker and number 10 in the world. Taj Hotels Resorts and Palaces is India's biggest luxury hotel group. Tata Power is the country's largest private electricity company. Tata Global Beverages is the world's second-largest maker of branded tea.

Tata operates in over 80 countries and employs about 600,000 people. Overall, the group earned 7 trillion rupees, or US$108 billion, in revenues in 2015 and 45 billion rupees in profits. Nearly 60 per cent of its revenue comes from outside India.

Apple Inc.

Apple Inc. is one of the richest corporations in the world, valued at over US$900 billion in 2018. However, the Apple supply chain has received much criticism on account of human rights and environmental and ethical issues in China.

For the manufacture of the iPhone, Apple has some 785 suppliers in over 230 countries worldwide—349 of them in China. In its Supplier Code of Conduct, Apple states that "suppliers are required to provide safe working conditions, treat workers with dignity and respect, act fairly and ethically, and use environmentally responsible practices wherever they make products or perform services for Apple".

Concept link

PLACES: The dynamic flow of goods, capital and services depends on the role of governments and corporations in facilitating the transfer of these items. Places can become more attractive over time for FDI, which boosts the power of corporations and can also bring possibilities to a place in terms of economic development. The majority of the global flows are legal, but illegal goods and services are also part of a global network.

Test yourself

4.16 Briefly **explain** three reasons why companies such as the Tata Group and Apple Inc. invest in other countries.   [2+2+2]

4.3 HUMAN AND PHYSICAL INFLUENCES ON GLOBAL INTERACTIONS

You should be able to show how political, technological and physical processes influence global interactions:

✔ Political factors that affect global interactions:

  ✔ multi-governmental organizations (MGOs) and free trade zones (FTZs);

  ✔ economic migration controls and rules;

✔ Our "shrinking world" and the forces driving technological innovation:

  ✔ changing global data flow patterns and trends;

  ✔ transport developments over time;

  ✔ patterns and trends in communication infrastructure and use;

✔ The influence of the physical environment on global interactions:

  ✔ natural resource availability;

  ✔ the potentially limiting effect of geographic isolation, at varying scales.

> • **Free trade zone (FTZ)** – a relatively small-scale special economic zone in which goods may be imported and manufactured and re-exported without customs duty (tax). They are often located close to ports, airports and/or national boundaries to take advantage of location for trade.
>
> • **Multi-government organization (MGO)** – an organization consisting of several sovereign states.

Political factors that affect global interactions

Multi-governmental organizations (MGOs)

Until relatively recently, the world was becoming more connected. There were many advantages of this for investors, TNCs and nations, including:

- greater access to raw materials and natural resources

- access to new markets

- access to sources of cheap and/or skilled labour

- the ability to achieve competitive advantage through increased efficiency, greater productivity or cost reduction.

However, since 2010, there has been a rise in nationalism (for example, the UK's "Brexit" vote), protectionism (for example, President Trump's policies in the USA), an increase in tariffs (the USA and China) and a stepping back from global interactions (for example, the Trans-Pacific Partnership).

The Trans-Pacific Partnership (TPP) has been replaced by the Comprehensive and Progressive Agreement for Trans-Pacific Partnership (CPTPP). This is now the world's third-largest trade agreement and represents 13.4% of global gross domestic product.

In 2018, the UK expressed an interest in joining CPTPP (as did the USA).

> **Concept link**
>
> **PROCESSES:** Human and physical factors determine the level of interaction between places. These factors include government policy, the availability of raw materials and the technological infrastructure which change over time and facilitate the interactions between places. Conversely, these factors can present challenges which may result in the implementation of new measures, such as the restriction the freedom of movement of people, capital and information, or the development of energy production which may reduce the level of interaction between countries.

Test yourself

4.17 Outline the advantages and disadvantages of multi-governmental organizations, such as the CPTPP. [3+3]

4.18 State why the UK's interest in joining the CPTPP is considered strange by many politicians. [2]

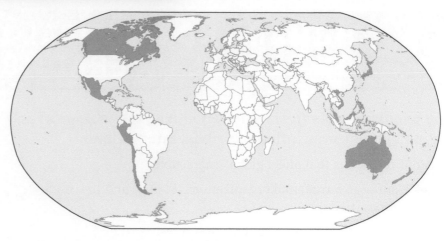

▲ Figure 4.3.1. Signatories to the CPTPP, in green

Free trade zones

A free trade zone (FTZ) is a type of special economic zone in which goods may be imported and manufactured and re-exported without customs duty (tax). They are often located close to ports, airports and/or national boundaries to take advantage of the location for trade. Originally free trade zones offered warehousing, storage and distribution facilities for trade, trans-shipment and re-export operations, but increasingly they focus on service industries including software, financial services, research and back-office operations.

Test yourself

4.19 Briefly **explain** one advantage and one disadvantage of free trade zones. [2+2]

An export processing zone (EPZ) is a small area, usually less than 300 ha, that specializes in manufacturing for export. In China, 70% of goods in EPZs must be exported, whereas in an FTZ there is no quota.

Economic migration controls and rules

Illegal immigration to the USA refers to foreign nationals voluntarily residing in the country in violation of US immigration and nationality laws. Illegal immigration carries a civil penalty. Punishment can include fines, imprisonment and deportation.

Assessment tip

Keep referring back to the question so that you make sure your answer is relevant and gains as many marks as possible.

It is estimated that there are around 12 million illegal immigrants in the USA.

Around 5.5 million migrants have over-stayed their visas whereas about 7 million crossed the border illegally.

Content link

Connect this information with the aspects of migration discussed in units 1.1 and 1.2.

Illegal migration over the US–Mexico border is concentrated around big border cities such as El Paso in Texas and San Diego in California, which have extensive border fencing and enhanced border patrols. Stricter enforcement of the border in cities has failed to curb illegal immigration, instead pushing the flow into more remote regions.

Test yourself

4.20 Analyse reasons why some countries attempt to control migration. [3+3]

In 2007, the US Congress approved a plan calling for more fencing along the Mexican border, with funds for approximately 700 miles (1,100 km) of new fencing. In 2017, US President Donald Trump called for the wall to be completed—and to be paid for by Mexico.

Our "shrinking world" and the forces driving technological innovation

Changing global data flow patterns and trends

The development and expansion of innovative technology, including the internet, e-mail and smartphones, has changed the ways people interact, communicate and share. Individuals and businesses buy, sell and communicate via the internet every day. The transfer of information, or data, is referred to as data flows.

Some countries are concerned about the free flow of data, and have placed barriers to restrict such flows. In trade agreements, governments can restrict the data that crosses borders. However, unnecessary barriers to cross-border data flows create considerable obstacles to global trade. With increased restrictions on data flows, global trade may suffer.

About one-eighth of the global goods trade is carried out by international e-commerce, and about half of the world's traded services are already digitized. Digitization allows the instantaneous exchange of virtual goods such as e-books, apps, online games, MP3 music files and streaming services, and software. For example, when Netflix expanded its business model from mailing DVDs to online streaming, it increased its market to over 190 countries. Digital platforms allow companies to reach beyond the constraints of a small local market and reach a global audience.

Although more countries are participating in digital networks, global flows are concentrated among a comparatively small number of users. The USA, Europe and Singapore are at the centre of the world's digital networks.

Transport developments over time

"Time–space convergence" refers to the ways in which the time required to "travel" from one place to another is decreasing. It suggests that places are getting closer together—not in a physical sense but in the time needed to travel between them. For example, between 1500 and 1840, the best average speed for horse-drawn coaches and ships was just 10 mph. This meant it took a long time to travel between places. By the 1960s, passenger jets could travel at speeds of between 500 and 700 mph. Thus, the time needed to travel between faraway places was much reduced. Finally, in the modern era, information can be sent around the world by the internet in a matter of seconds.

The size of container ships has increased over 14-fold since the 1960s. The largest container ships are now over 19,000 TEUs (twenty-foot equivalent units). The advantages are that more goods can be transported, and the larger size reduces the cost of transport per unit transported. However, there are relatively few ports that can accommodate these ships. Moreover, some canals may be too small— the Panama Canal is being widened so that it can allow ships of up to 12,000 TEUs to pass. Some seas are too shallow to allow boats above a certain size to pass—the Straits of Malacca linking the Indian Ocean and the Pacific Ocean is a good example.

Patterns and trends in communication infrastructure and use

Information and communication technologies (ICTs) offer increased potential for advancing progress towards economic and social development objectives.

Electronic commerce (e-commerce) is growing in all parts of the world. But many developing countries remain relatively unprepared for the shift from offline to online trade. Less than 5% of the population in most of these economies buy goods and services online.

> **Assessment tip**
>
> Information about data flows becomes obsolete very quickly. It is a good idea to gather information throughout your IB course, so that you may show how it changes over time.

▲ Figure 4.3.2. Developments in container ships have had a major impact on economic geography

Test yourself

4.21 Identify the main winners and losers as a result of
(a) changing data flows and
(b) transport developments. [1+1]

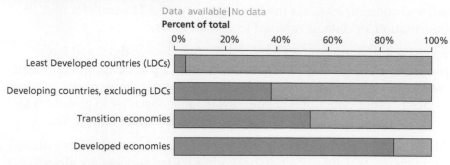

▲ Figure 4.3.3. Share of countries having data on ICT use by businesses, by development group, 2003–2016

Test yourself

4.22 Using figure 4.3.3, **examine** the extent to which digitization enabled equal access to the benefits of globalization. [8]

The digital economy is bringing new risks along with opportunities, and emerging technologies such as advanced robotics, artificial intelligence, the Internet of Things, cloud computing and 3D printing are already disrupting economies and societies.

The influence of the physical environment on global interactions

Natural resource availability

Natural resource availability is another important determinant of physical trade patterns, as it has a high correlation to physical trade balances. Since 1990, 10% of the resource-rich countries (which in absolute numbers is made up of 15 countries) have been net suppliers of materials to global markets. There are a few remarkable exceptions. Some of the most resource-rich countries, such as China, the USA and India, were net importers, and some resource poor countries, like Guyana or Latvia, were net exporters of materials.

Content link

Resource availability connects with resource security, which is discussed in unit 3.2.

The countries with the highest proportion of resources are mainly large countries, such as Russia, Canada, USA, China, Saudi Arabia, India, Brazil and Australia. Countries with the least amount of natural resources include a number of Sub-Saharan African countries, such as Mali, Mauritania, Namibia and Botswana.

The potentially limiting effect of geographic isolation, at varying scales

Isolation from world markets can be a major limiting factor. Peripheral areas have increased transport costs and times which may reduce access to markets. Countries that are landlocked may also have to pay substantial tariffs to export their goods. For example, South Sudan has to pay either Sudan or Kenya to get its oil to the coast. Landlocked countries may also have to pay for the use of another country's air space.

Access to market has long been a theme in geography, and models of agriculture and industry have considered location in relation to markets and labour. This occurs at a local scale as well as at a regional and international scale. For example, there is a global core (EU–North America–Japan) and periphery (mostly in Sub-Saharan Africa), but there are also national and local cores and peripheries.

QUESTION PRACTICE

a) Analyse how the global influence of different superpower states has changed over time. [12]

b) Examine reasons why it is hard to measure the real size of different global flows (such as migration or trade). [16]

How do I approach these questions?

a) The command term is "analyse": break down the subject of the question in order to bring out the essential elements or structure. In this case, you should show how superpowers have changed, and why they have changed. To achieve the most marks you should provide examples to back up the points you make.

The focus should be on changing superpowers: which countries, when did they become superpowers, and what made them a superpower.

Good answers are likely to say why some superpowers have declined.

b) The command term is "examine": consider an argument or a concept in a way that uncovers the assumptions and interrelationships of the issue. In this question you should **examine** the characteristics of different global flows and the resulting attempts to measure the extent of these flows.

SAMPLE STUDENT ANSWER

a) A superpower is a country that is able to influence policy on a worldwide scale. A superpower may have economic, cultural, military and geographical influence on the rest of the world, as seen, for example, in the USA and, increasingly, in China. Superpowers generally have a large population, land area, economy and access to resources. Superpowers can grow and decline – the UK was a superpower during the second half of the nineteenth century and first half of the twentieth century. However, as its former colonies took their independence after the Second World War, the UK declined as a superpower.

Following the break-up of the Soviet Union in 1991, the USA was for a time the world's only superpower. However, the USA has not always been able to achieve its goals – military involvement in Afghanistan has not been entirely successful, and many critics argue that the USA under President Trump has become isolationist and inward-looking, and that the USA is losing its superpower status.

The growth of China, economically, is rivalling that of the USA. It has also been argued that the growth of the European Union, a group of 28 countries (at the time of writing, the UK has not left the EU), has also been described as a superpower, but internal differences between member countries reduce its cohesion.

Since 1992 China's economic growth has been substantial. China is now a major trading partner with the USA, Japan, South Korea and the EU. However, President Trump has threatened a trade war with China. China's economic structure remains different to that of the USA – it has around 28% employed in agriculture, 29% in manufacturing and 43% in services, compared with 2%, 23% and 75% respectively in the USA. China is the world's leading manufacturing country. However, its PPP is $16,600 compared with $59,500 of the USA.

- ▲ Correct definition, valid point
- ▲ Further development—another valid point
- ▼ Correct point but very general—would be better with support
- ▼ Reasons for its growth would have been good—should refer back to the question and tie the case study to the question directly
- ▲ A valid description of the decline of the UK as a superpower
- ▼ Some dates or more detail would gain credit here
- ▲ Good contemporary point—could give some detail on the US's isolationist position
- ▼ Changing nature of superpowers implied but not explicit
- ▼ Needs to refer back to the question to make it relevant
- ▲ Focus on changing superpowers—the growth of China
- ▼ Could give data or more detail
- ▲ Good supporting data which shows the contrast between China and the USA

▲ Shows some contrast

▲ Definition of soft power

▲ Develops idea of soft power

▲ This is a good section on cultural and political influences

▲ A balanced approach showing that the USA is doing better in some areas, eg soft power, but that China is taking the lead on climate change

The USA also has more "soft power" than China. "Soft power" is the ability to change individuals, communities and nations without using force or coercion. Many countries achieve soft power through their culture, political values and foreign policies such as aid and investment. For example, the USA favours democracy and human rights. The distribution of American films and TV programmes, such as Friends, has led to the Americanization of other cultures and languages. On the other hand, the USA has not been at the front of leading actions to tackle environmental issues such as climate change. The Chinese President Xi Jinping has vowed to protect the Paris Agreement, which aims to limit climate change and fossil fuel emissions.

Overall, a very good essay. Makes many good points with support. However, it could refer back to the question more directly.

Marks 9/12

▲ Identifies KOF Index

▲ Valid points about complexity

▲ Pragmatic point

▲ Valid point about size of most globalized countries against some that are less globalized

▲ Contemporary issues—all valid

▲ Good exemplification

b) There are many indices that try to show the scale of global flows. One of these is the KOF index, which examines social, economic and political flows. For example, it includes long-distance flows of goods, capital and services, the spread of ideas, information and people. Other indices such as the EY Globalisation Index look at openness to trade, capital flows, exchange of technology and ideas. These show just how complex global flows can be. In addition, not all of the data are readily available e.g. data from the informal economy, remittances, illegally traded goods and people smuggling. Although it is easy to criticise KOF on the choice of some of its indices, it is making an attempt to explain a highly complex situation with the data that are available.

Moreover, some of the more globalized countries, with large volumes of cross-border 'traffic' are small countries e.g. Singapore, Netherlands and Sweden, whose volume of global flows may be relatively small. In contrast, large countries such as the USA and Russia may have large internal flows which include larger distances than for the small countries, or for countries within the EU trading bloc.

Moreover, many global flows are decreasing in size/distance in the current era of economic protectionism, anti-migration policies and resource nationalism. For example, some off-shoring and FDI has been replaced with re-shoring, so in many cases, countries and certain individuals within them, are choosing to be more isolated and self-centred e.g. America First and Brexit.

Many financial flows are difficult to measure. These include remittances. Some money is sent directly from migrants to their families, and so is not recorded. In other instances, some wealthy people have been lodging money in offshore accounts to avoid paying tax. Other individuals may set up companies into which they get paid, so as to avoid paying so much tax.

▲ Valid point—flows made by individuals

▼ Going slightly off course here with tax avoidance

There are many illegal flows, including those of people, narcotics and counterfeit goods. These are believed to be large-scale but are difficult to measure as they are illegal and participants go at great length not to be detected. Trafficking of people has a global scope. It can be domestic or international. Most trafficking is within a region, and victims are generally from a poorer country and trafficked to a richer country. Counterfeit goods are also a worldwide problem, and their distribution and trade is often a highly organized system. The trade in counterfeit goods has been linked to money laundering, illicit drugs and corruption. The flow of drugs involves cultivation, manufacture, distribution and sales of drugs.

▲ Valid points

▲ Some detail

▼ A bit too much in terms of definitions rather than answering the question

Over 500 different flows of trafficking were discovered between 2012 and 2014. For example, victims from Sub-Saharan Africa were mainly found in Africa, the Middle East and Western and Southern Europe. Some 42 per cent of detected victims between 2012 and 2014 were trafficked domestically. Since 2004, the profile of detected trafficking victims has changed. Although most detected victims are still women, children and men now make up larger shares of the total number of victims than they did in 2004. In 2014, children accounted for 28 per cent of detected victims, and men 21 per cent. The share of victims who are trafficked for forced labour has also increased. About 4 in 10 victims detected between 2012 and 2014 were trafficked for forced labour, and out of these victims, nearly two-thirds were men.

▲ Supporting statistic

▼ Detailed account but no indication of size

▼ Again, a detailed account but no indication of size

So overall, we can see that it is very difficult to measure global flows because there are so many of them, and some are hidden. Moreover, we are comparing different sized countries, some legal and illegal flows, and political and economic changes in the way the world operates. Global flows are still very important although they may be changing.

Generally a clear and focused answer. Some valid points with exemplification, but also some information included not directly related to the question.

Marks 10/16

5 HUMAN DEVELOPMENT AND DIVERSITY (HL ONLY)

This unit looks at different aspects of human development and inequalities. It examines ways in which human development can be improved. Differences in culture are considered, as are ways in which culture changes. In some cases, resistance to cultural change and global interactions can be observed.

You should be able to show:

✔ ways of supporting the **processes** of human development;

✔ how global interactions bring cultural influences and changes to **places**;

✔ the varying **power** of local places and actors to resist or accept change.

5.1 DEVELOPMENT OPPORTUNITIES

- **Gender** – the array of socially constructed roles and relationships, personality traits, attitudes, behaviours, values, relative power and influence that society ascribes to the two sexes on a differential basis.

- **Sex** – the biological characteristics that define humans as female or male.

- **Gender equality** – the concept that all human beings, both women and men, are free to develop their personal abilities and make choices without the limitations set by stereotypes, rigid gender roles or prejudices.

- **Gender equity** – means that women and men are treated fairly according to their respective needs.

- **Empowerment** – means that women and men can take control of their lives, set their own agendas, gain skills, increase self-confidence, solve problems and develop self-reliance.

- **Sustainable development goals (SDGs)** – a set of goals developed by the United Nations to end poverty, promote peace and to protect the planet.

You should be able to show ways of supporting the processes of human development:

✔ The multidimensional process of human development and ways to measure it:

   ✔ UN Sustainable Development Goals (SDGs) criteria;

   ✔ Validity and reliability of development indicators and indices, including the Human Development Index (HDI) and Gender Inequality Index (GII);

   ✔ Empowering women and indigenous or minority groups;

   ✔ Detailed illustrative examples of affirmative action to close the development gap;

✔ The importance of social entrepreneurship approaches for human development:

   ✔ The work of microfinance organizations and their networks;

   ✔ Alternative trading networks such as fair trade;

   ✔ TNC corporate social responsibility frameworks and global agreements.

The multidimensional process of human development and ways to measure it

UN Sustainable Development Goals criteria

Goal 1: Eradicate poverty and promote prosperity in a changing world

Goal 2: End hunger, achieve food security and improved nutrition, and promote sustainable food production

Goal 3: Ensure healthy lives and promote well-being for all at all ages

Goal 4: Ensure inclusive and equitable education and promote lifelong learning opportunities for all

Goal 5: Achieve gender equality and empower all women and girls

Goal 6: Ensure availability and sustainable management of water and sanitation for all

Goal 7: Ensure access to affordable, reliable, sustainable and modern energy for all

Goal 8: Promote sustained, inclusive and sustainable economic growth, full and productive employment and decent work for all

Goal 9: Build resilient infrastructure, promote inclusive and sustainable industrialization, and foster innovation

Goal 10: Reduce inequality within and among countries

Goal 11: Make cities and human settlements inclusive, safe, resilient and sustainable

Goal 12: Ensure sustainable consumption and production patterns

Goal 13: Take urgent action to combat climate change and its impacts

Goal 14: Conserve and sustainably use the oceans, seas and marine resources for sustainable development

Goal 15: Protect, restore and promote sustainable use of terrestrial ecosystems, sustainably manage forests, combat desertification, halt and reverse land degradation, and halt biodiversity loss

Goal 16: Promote peaceful and inclusive societies for sustainable development, provide access to justice for all, and build effective, accountable and inclusive institutions at all levels

Goal 17: Strengthen the means of implementation and revitalize the Global Partnership for Sustainable Development

Validity and reliability of development indicators and indices

The Human Development Index (HDI) is a composite measure of development. It includes three basic components of human development:

- Longevity (life expectancy)

- Education index—mean years of schooling or expected years of schooling

- Standard of living—income adjusted to local cost of living, that is, purchasing power.

The United Nations has encouraged the use of the HDI as it is more holistic than single indicators such as Gross National Income (GNI) per head. (GNI was previously known as gross national product.) It is a composite index so that the importance of any one factor is reduced.

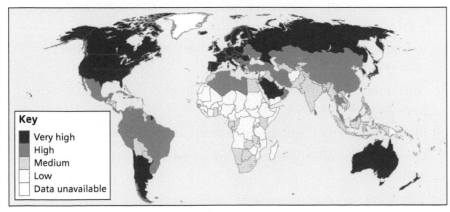

▲ **Figure 5.1.1.** The Human Development Index, 2017 (based on 2015 and 2016 data)

Key
- Very high
- High
- Medium
- Low
- Data unavailable

Content link

The 17 SDGs relate to many aspects of the IB Geography Course, for example, oceans (goal 14), deserts (goal 15), urban environments (goal 11), food and health (goal 2) and freshwater (goal 6). Some are related indirectly, such as tourism (goal 16).

Test yourself

5.1 Suggest a definition for the terms "inclusive" (SDG 4, 8, 9, 11, 16), "foster" (SDG 9) and LDCs and SIDS (SDG 9, 10).　[1+1+1]

5.2 Identify the environmental issues covered by the SDGs.　[2]

5.3 Describe ways in which gender inequalities remain a problem.　[4]

Concept link

PROCESSES: Composite, multidimensional indices are regarded as the most appropriate methods of measuring the success of direct and indirect human development processes.

>> Assessment tip

Try to have some data for the HDI of contrasting countries, that is, include examples of HICs, MICs and LICs.

▼ Table 5.1.1. Human Development Index for the United Arab Emirates, 2015

HDI (value)	Life expectancy at birth (years)	Expected years of schooling (years)	Mean years of schooling	Gross national income (GNI) per capita PPP* (US$)	GNI per capita rank minus HDI rank
0.840	77.1	13.3	9.5	69,200	35

*PPP = power purchasing parity
Source of data: Human Development Report (2016)

The Gender Inequality Index (GII) measures gender inequalities in three aspects of human development:

- Reproductive health—as measured by maternal mortality ratio and adolescent birth rates

- Gender empowerment—as measured through the proportion of parliamentary seats held by women, and the proportion of adult females and males aged over 25 years with some experience of secondary school

- Economic status—as measured by labour force participation by males and females aged 15 and over.

▼ Table 5.1.2. The components of the gender inequality for selected countries

HDI Rank	Country	GII		Maternal mortality ratio (deaths per 100,000 live births)	Adolescent birth rate (births per 1,000 women aged 15–19)	Share of seats in parliament (%) held by women	Population with at least some secondary education (% ages 25 and older)		Labour force participation rate (% ages 15 and older)	
		Value	Rank				Female	Male	Female	Male
		2015	2015	2015	2015	2015	2005–15	2005–15	2015	2015
42	United Arab Emirates	0.232	46	6	29.7	22.5	77.4	64.5	41.9	91.6
186	Chad	0.695	157	856	133.5	14.9	1.7	9.9	64.0	79.3
187	Niger	0.695	157	553	202.4	13.3	3.6	8.4	40.2	89.4

Source of data: Human Development Report (2016)

Empowering women

UN Women (full title: the United Nations Entity for Gender Equality and the Empowerment of Women) has seven principles:

1. Establish high-level corporate leadership for gender equality

2. Treat all women and men fairly at work—respect and support human rights and nondiscrimination

3. Ensure the health, safety and well-being of all women and men workers

4. Promote education, training and professional development for women

5. Implement enterprise development, supply chain and marketing practices that empower women

6. Promote equality through community initiatives and advocacy

7. Measure and publicly report on progress to achieve gender equality.

The following lists outline some of the inequalities facing women today.

Women at risk:

- Women represent 70% of the world's poor.

- In the developing world, the percentage of land owned by women is less than 2%.
- An estimated 72% of the world's 33 million refugees are women and children.

Violence against women:

- The most common form of violence experienced by women globally is physical violence inflicted by an intimate partner.

Women and the business case:

- In Asia, the gender equality gap in employment rates for women cost US$47 billion annually through lost productivity—about 45% of women remained outside the workplace compared to 19% of men.

Women and education:

- About two-thirds of the estimated 776 million adults (or 16% of the world's adult population) who lack basic literacy skills are women.

Empowering women in Colombia

During the armed conflicts in Colombia (mid-1960s to 2013), women became mobilized and they have influenced new laws and policies. Colombia now has more women in decision-making positions than ever before: by 2011, 32% of cabinet members were women, up from just 12% in 1998. Girls' enrolment in secondary and tertiary education exceeds that of boys', and women's participation in the labour force increased from 30% in 1990 to nearly 43% in 2017.

Empowering indigenous peoples or minority groups

The Millennium Development Goals (MDG) Achievement Fund is working in 20 countries around the world to help indigenous people improve their health, preserve their cultures, protect their rights and boost their livelihoods. There are some 370 million indigenous people around the globe, constituting just 5% of the world's population but 15% of its poor. Largely marginalized and isolated, indigenous people are in worse health than the general population, and are much less able to influence and participate in political and economic processes that affect them.

In Panama, for example, the UN established a sustainable rural development project for the Ngäbe-Buglé Territory to restore land rights. It provided financial and technical support to the communities involved, and improved local leaders' planning and administrative skills.

Minority groups may also include immigrants, refugees and ethnic minority groups in a country (for example, the Rohingya people in Myanmar), who also experience discrimination and inequalities.

The importance of social entrepreneurship approaches for human development

Microfinance

In 1983, Muhammad Yunus launched the Grameen Bank in Bangladesh to allow the poor to establish microbusinesses and improve their lives. The idea has since spread across the developing world.

However, there is a downside. For example, a big expansion of microcredit in India's Andhra Pradesh province caused a crisis in 2010, when the lenders were blamed for an increase in suicides by farmers who could not pay back their loans.

Test yourself

5.8 Outline ways in which women are disadvantaged relative to men. [6]

5.9 Suggest two ways in which gender inequality could be reduced. [2+2]

>> **Assessment tip**

Make sure that you read the question carefully. Question 5.8 asks you for "ways" (plural) and Question 5.9 asks you for "two ways". So, in both parts you need to provide two (or more) examples.

>> **Assessment tip**

When answering questions that ask you to "suggest" or "explain", you should always try to use examples in your answers as support for the point that you make, and/or develop the explanation with cause–effect type comments.

Content link

Efforts to empower Aboriginal people are discussed in unit 6.3.

Test yourself

5.10 Suggest how the needs for indigenous people differ from those of non-indigenous people. [4]

5.11 Outline ways in which indigenous people are disadvantaged relative to non-indigenous people. [2+2]

Part of the problem is that microfinance is very hard to provide on a large scale. Typical annual percentage interest rates are in the region of 20–40%, cheaper than the traditional local moneylender or pawnbroker but still expensive.

Alternative trading networks such as fair trade

Fair trade can be defined as trade that attempts to be socially, economically and environmentally responsible. It is trade in which companies take responsibility for the wider impact of their business.

▼ Table 5.1.3. Fair trade pros and cons

Pros	Cons
It pays higher wages to its workers.	There is a limited customer base for fair trade products.
It considers worker safety.	There are fewer fair trade products to choose from than non-fair trade products.
There is no discrimination.	Fair trade costs may increase due to administrative costs.
It tries to eliminate the need for child labour.	Most large buyers are not attracted to the fair trade model.
Organic methods are often used to produce fair trade products.	The demand for fair trade products reduces in times of recession.

TNC corporate social responsibility frameworks and global agreements

Corporate social responsibility (CSR) began to appear in the 1950s, when business leaders and academics identified the significance of business decisions in the context of community welfare.

Improving and reporting on social (and environmental) performance makes good business sense for any corporation. Socially irresponsible behaviour could result in declining customer loyalty and reduced sales. Adverse publicity about social conditions and environmental impacts may damage a corporation's reputation.

For example, throughout the 1990s Nike was targeted by anti-globalization activists for allowing its suppliers in LICs to abuse and exploit workers. Following an article published in *Life* magazine with the heading "Six Cents an Hour" and a picture of a Pakistani boy sewing a Nike football, Nike became synonymous with slave wages, forced overtime and abuse. Nike gradually adopted an expensive and extensive system for monitoring and remedying factory conditions in its supply chain.

To achieve effective CSR, companies should address the following areas:

- Ways to minimize percentage of employee turnover;
- Ways to minimize salary inequalities;
- Ways to provide daycare services for employees with young children;
- Ways to provide flexible working hours or working hours that promote a work–life balance;
- Ways to improve the female:male ratio in managerial roles;
- Ways to ensure human rights are observed by suppliers.

>> **Assessment tip**

Don't confuse fair trade and free trade. Fair trade tries to be socially, economically and environmentally responsible whereas free trade allows trade between two countries and can be quite exploitative.

Test yourself

5.12 Outline one advantage and one disadvantage of microfinance. [2+2]

5.13 Describe one advantage and one disadvantage of fair trade. [2+2]

5.14 Briefly **explain** the advantages of corporate social responsibility (CSR). [3]

The UN's International Labour Organization is the specialist UN agency that deals with labour problems, especially those in relation to international labour standards, social protection and work opportunities for all. It has 187 members. Countries not belonging to the ILO include North Korea and Eritrea.

5.2 CHANGING IDENTITIES AND CULTURES

You should be able to show how global interactions bring cultural influences and changes to places:

✔ The global spectrum of cultural traits, ethnicities and identities, and ways in which the spectrum of diversity is widening or narrowing at different scales;

✔ The effects of global interactions on cultural diversity in different places:

  ✔ the diffusion of cultural traits, and cultural imperialism;

  ✔ glocalization of branded commodities, and cultural hybridity;

  ✔ cultural landscape changes in the built environment;

✔ How diasporas influence cultural diversity and identity at both global and local scales;

  ✔ Case study of a global diaspora population and its cultures.

The global spectrum of cultural traits, ethnicities and identities

Culture gives people a sense of community and belonging. Until recently, cultures had been considered spatially bounded, although the rise (and fall) of some empires led to a two-way spread of certain cultural traits (sport and food, for example). Contemporary globalization has led to ties between distant and disparate places, ideas and symbols.

Advances in information technology during the 20th century have enabled cultural symbols (such as music, images and text) to flow around the world faster than ever before. Some argue that this has led to a new "global consumer culture" based on the diffusion of Western culture, around commodities such as McDonald's and Coca-Cola. Such symbols of Western culture have become widespread around the world, but the idea that they have shaped a new global consumer culture is hotly debated.

Current cultural globalization differs from cultures of the past in many ways such as:

• the scale of cultural exchanges (it is not a one-way process)

• the rise of TNCs in the culture industries (for example, film, music, TV)

• the rise of the business culture.

Globalization has led to the mixing, or hybridization, of culture rather than cultural imperialism. Westernization/Americanization of culture may be a powerful process, but it can be (and has been) rejected in many places.

• **Culture** – a system of shared meanings used by people who belong to the same community, group or nation, to help them interpret and make sense of the world. These systems of meanings include language, religion, custom and tradition, as well as ideas about "place".

• **Cultural diffusion** – the spread of cultural ideas, from their place of origin to other regions, groups or nations.

• **Cultural imperialism** – the practice of promoting the culture, values or language of one nation in another, less-powerful one.

• **Diaspora** – the forced or voluntary dispersal of any population sharing a common racial, ethnic or cultural identity, after leaving their settled territory and migrating to new areas.

• **Glocalization** – the adaptation of a product or service to the locality or culture for which it is marketed.

• **Homogenization** – the process by which features become increasingly similar or uniform.

Test yourself

5.15 Explain why some cities would choose to be similar to other cities. [6]

The effects of global interactions on cultural diversity in different places

Culture represents the systems of shared meanings that people from the same community, group or nation use to help them make sense of the world. These systems of meanings include language, religion, custom and tradition, and ideas about "place".

- Language: A number of languages have more than 100 million native speakers. These include English, Mandarin (Chinese), Spanish, Portuguese, Hindi, Arabic, Russian and Bengali. English has become one of the dominant world languages.

- Music: The production, distribution and consumption of music have a particular geography. Transnational corporations have a powerful influence over the global music industry, with artists from the USA and UK dominating the global popular music market.

- Television: Until the early 1990s, television programmes tended to be produced primarily for domestic audiences within national boundaries, and could be subjected to rigorous governmental control.

The diffusion of cultural traits, and cultural imperialism

New technologies, such as the internet and satellite communications, mean that the world is becoming more interconnected. The increased speed of transport and communications, the increasing interactions between economies and cultures, the growth of international migration, and the power of global financial markets are among the factors that have changed everyday lives in recent decades.

Global cultural imperialism today has resulted from economic forces, for example, when the dominant culture (usually the USA) captures markets for its commodities and thereby gains influence and control over the popular culture of other countries. Below are some examples.

- Language: There are around 6,000 languages in the world, but this figure may reduce to 3,000 by 2100. English is becoming *the* world language due to the global influence of the USA. Although Mandarin is more widely spoken as a first language, the total number of English speakers, if second-language speakers are taken into account, is close to 1 billion. English is the medium of communication in many important fields, including the internet.

- Media: National media systems are being superseded by global media complexes. Around 20 to 30 large TNCs dominate the global entertainment and media industry, mainly from the West, and most of which are from the USA. These include giants such as Time-Warner, Disney, News Corporation, Universal Studios and the BBC.

Glocalization of branded commodities, and cultural hybridity

Cultural hybridity is the merging of previously separate cultural traits. For example, there is a phenomenon in the USA that involves the merging of the Spanish language and the English language to produce "spanglish". Roughly one in every seven US residents is of Latino origin. In US states that are close to the US–Mexico border, some radio and TV stations use "code switching", that is, words and phrases in Spanish are inserted haphazardly into English sentences, and vice versa.

▲ **Figure 5.2.1.** This McDonalds restaurant in Dubai is an example of cultural globalization

Test yourself

5.16 Describe the trend in **(a)** global languages and **(b)** global media since the 1970s.   [2+2]

5.17 Briefly **explain** how and why these trends have happened.   [2+2]

Another example of cultural hybridity is the development of Tex-Mex food. Tex-Mex was developed in the Rio Grande valley in Texas, which had a large Mexican population and a large cattle industry. In the 1870s, a group of Hispanic women started serving chili con carne to Americans. Later they added beans, rice and sour cream or grated cheese, and so began the "Mexican menu". Later, Mexican migrants in Texas developed tortillas, nachos, fajitas and the breakfast taco. Tex-Mex has been described as America's oldest regional cuisine, and it continues to evolve today.

Cultural landscape changes in the built environment

The evolution of uniform urban landscapes is the result of a variety of factors:

- improvements in communications technology, such as television- and internet-based technology, so that people in one city are aware of opportunities and trends in other cities

- increased international migration and the spread of ideas and cultures

- time–space convergence, which allows faster interactions between places

- the desire of global brands (TNCs) such as Apple, Google and Starbucks to reach new markets

- improvements in standards of living and aspirations to be part of a global network of urban centres

- globalization of economic activity, culture (art, media, sport and leisure activities) and political activity

- attempts to create smart cities.

> ### Concept link 🔗
>
> **PLACES:** There are different processes that can diffuse culture from one place to another. Places tend to resist, adopt or adapt to the introduction of new traits, which subsequently retains the cultural makeup of a place, alters it or perhaps completely changes the identity.

Many urban landscapes in different countries today look very similar. Tall towers are a feature of many cities. Industrial estates and science parks are increasingly globalized as TNCs outsource their activities to access cheap labour, vital raw materials and potential markets. Many cities have pedestrianized shopping centres, open markets and out-of-town supermarkets.

All urban areas have something in common. All have something unique. Urban areas that are less Westernized or less globalized might be expected to be more different from one another than those that are more globalized. For example, the urban landscape of Bandar Seri Begawan, Brunei, is dominated by a mosque. A large proportion of people live in Bandar's "water village", yet even there evidence of Western culture can be found.

▲ **Figure 5.2.2.** The urban landscape of Kampong Ayer, Bandar Seri Begawan's "water village"

How diasporas influence cultural diversity and identity at both global and local scales

The term diaspora refers to any dispersal of a population formerly concentrated in one place. Examples include:

- professional and business diasporas; for instance, Samsung Electronics' UK division was originally based in New Walden, and as a result many South Koreans moved to this area.

> ### ⟩⟩ Assessment tip
>
> Remember, in your discussion of cities, that no two cities are the same. Every city is unique in certain aspects, although there are certain aspects that may be similar.

- cultural diasporas, such as the movement of migrants of African descent from the Caribbean to the UK after World War II. The British government encouraged immigration from Commonwealth countries to fill shortages in the workforce.

Diasporas may bring their culture with them. Some may marry people from other population groups, and so cultures may be mixed, or hybridized.

Case study: The Chinese diaspora

Approximately 40 million people of Chinese origin live in sizeable numbers in at least 20 countries. Large concentrations are found in Singapore (2.6 million), Indonesia (7.6 million), Malaysia (6.2 million), Thailand (7 million) and the USA (3.4 million). Historically, Chinese migration began in the 10th century with the expansion of maritime trade. During periods of colonialism, large numbers of Chinese moved into Singapore and Mauritius—the latter encouraged by the French. With globalization, Chinese migration for professional and business reasons has increased.

> **Test yourself**
>
> **5.18 Describe** ways in which the built environment is becoming more similar around the world. [5]
>
> **5.19** Briefly **explain** why the built environment is becoming more similar around the world. [4]

5.3 LOCAL RESPONSES TO GLOBAL INTERACTIONS

You should be able to show the varying power of local places and actors to resist or accept change:

- ✔ Local and civil society resistance to global interactions:
  - ✔ rejection of globalized production, including campaigns against TNCs and in favour of local sourcing of food and goods by citizens;
  - ✔ rise of anti-immigration movements;
- ✔ Geopolitical constraints on global interactions:
  - ✔ government and militia controls on personal freedoms to participate in global interactions;
  - ✔ national trade restrictions, including protectionism and resource nationalism;
- ✔ The role of civil society in promoting international-mindedness and participating in global interactions, including social media use and campaigning for internet freedom;
  - ✔ Two detailed examples of places where restricted freedoms have been challenged.

> • **Civil society** – any organization or movement that works in the area between the household, the private sector and the state to negotiate matters of public concern. Civil societies include non-governmental organizations (NGOs), community groups, trade unions, academic institutions and faith-based organizations.
>
> • **Resource nationalism** – when a country decides to take all or part of one or a number of natural resources under state ownership.
>
> • **Protectionism** – any economic policy that limits trade between countries in order to protect trade in the home country.

Local and civil society resistance to global interactions

Rejection of globalized production

According to Greenpeace, TNCs control global food production.

- Six corporations—Monsanto, DuPont, Dow, Syngenta, Bayer and BASF—control 75% of the world pesticides market.

- Factory farms now account for 72% of poultry production, 43% of egg production and 55% of pork production worldwide.

- Four corporations—ADM, Bunge, Cargill and Dreyfus—control more than 75% of the global grain trade. They

▲ Figure 5.3.1. Farmers' market in Woodstock, UK

overwhelmingly force commodity crops like corn and soy on local farmers at the expense of native crops.

Those who favour local production stress that it increases market access and sales for the producer, and that it improves consumer understanding of food production. Local production also provides consumers with fresher food, reduces food miles and has a smaller carbon footprint. It improves the local farming economy, and has a multiplier effect, for example, increased demand for fodder, vets and farm equipment.

The rise of anti-immigration movements

There are many reasons for the rise of anti-immigration groups. The main concerns of those opposed to immigration are the perceived threats over competition for jobs, and the cost of housing, education and health care. In some cases, notably in LICs and NICs, environmental issues may also be a concern as a result of rapid population growth. Some argue that certain immigrant groups isolate themselves from society and refuse to integrate into mainstream society. If migrants are unable to assimilate into society, they may form ghettos. Other perceived concerns include increased crime rates and the spread of infectious diseases.

Key
☐ Pro-migration   ☐ Anti-migration

Country	Pro-migration	Anti-migration
Germany	66%	29%
United Kingdom	52%	37%
United States	51%	41%
Spain	47%	46%
France	45%	52%
Poland	24%	52%
Greece	19%	70%
Italy	19%	69%

▲ **Figure 5.3.2.** Popular views of immigrants in selected countries

Source of data: Pew Research Center (2014)

Geopolitical constraints on global interactions

Government and militia controls on personal freedoms to participate in global interactions

Myanmar's government began a large-scale ethnic cleansing campaign against the Rohingya Muslim population in Rakhine State in August 2017. Over 650,000 Rohingya (out of a total of 1 million) fled to neighbouring Bangladesh to escape mass killings, sexual violence, arson and other abuses. Despite the election of the National League for Democracy (NLD) and its leader, Aung San Suu Kyi, the government has increasingly used repressive laws to control journalists and critics of the government/military.

Religious minorities, including Muslims, Christians and Hindus face persecution in a country that is nearly 90% Buddhist.

Human trafficking is a problem, especially in the north of Myanmar. According to a joint report by the John Hopkins School of Public Health and the Kachin Women's Association Thailand, 7500 women and girls have been trafficked from Myanmar into forced marriages in China over the last five years.

Meanwhile, some Western governments that had previously criticized Myanmar on its human rights record now see it as a land of

Test yourself

5.20 Outline the advantages of mass-produced food.   [4]

5.21 Describe the main characteristics of the farmers' market shown in figure 5.3.1.   [3]

5.22 Outline the disadvantages of reliance of farmers' markets.   [3]

Test yourself

5.23 Suggest reasons for the differences in views concerning immigrants into the countries shown in figure 5.3.2.   [3]

Content link
The displacement of the Rohingya people is related to the aspects of forced migration discussed in unit 1.2.

opportunity and reform. Investors have increased, and their dealings are largely through the NLD rather than civil society organizations (CSOs), who are now less able to raise international support than when Myanmar was under military dictatorship. The opportunities for CSOs in Myanmar appear to be reducing under the new NLD government. CSOs have experienced a major brain drain in recent years.

National trade restrictions, including protectionism and resource nationalism

Protectionism refers to any economic policy that limits trade between countries in order to protect trade in the home country.

> **Assessment tip**
>
> Try to keep up to date with contemporary changes in protectionism.
>
> Much protectionism is actually carried out by HICs.

Test yourself

5.24 Outline the characteristics of the countries with the highest level of protectionism. [3]

5.25 Describe the distribution of countries with a low level of protectionism. [3]

5.26 Examine the "arguments" for and against protectionism. [12]

5.27 Suggest reasons for the growth of nationalism since 2008. [4]

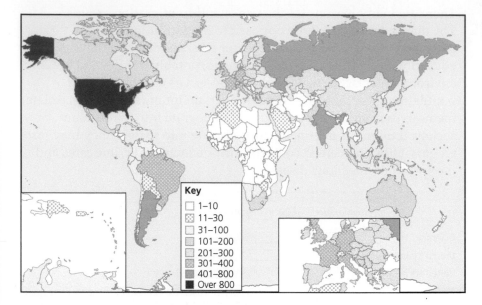

▲ **Figure 5.3.3.** Number of protectionist measures taken by different countries, 2008–2016

Source of data: Global Trade Alert

Resource nationalism

Resource nationalism is when a country decides to take all or part of its natural resources under state ownership. It affects both the host country and any TNC involved in the development of the resource in the first place. The main advantage for the host country is that they can receive more revenue from the development of the resource. However, they may lack the equipment, technology or finance to develop it fully, and so may not benefit as much as they might do. Where a TNC has been involved in the development of the resource in the first place, it is likely to make large profits. However, if the host government decides to nationalize the resource, the TNC may spend large amounts of money in research and development only to see its investment disappear. The host and the TNC rely on each other.

The role of civil society in promoting international mindedness and participation in global interactions

According to Freedom House, the internet is a crucial medium through which people can express themselves and share ideas. It is also an increasingly important tool through which democracy and human rights activists mobilize and advocate for political, social and economic reform. Some authoritarian states have devised ways to filter, monitor, obstruct and/or manipulate the openness of the internet.

Places where restricted freedoms have been challenged: The Arab Spring

The Arab Spring relates to the range of demonstrations, protests, riots and civil wars that spread through countries in the Middle East and

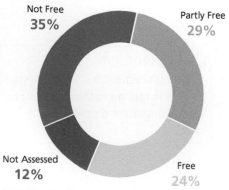

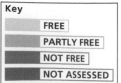

▲ **Figure 5.3.4.** Freedom of the internet in the world

North Africa after 2010. Most of the revolutions and protests were over by 2012, although the ongoing conflict in Syria is an important exception. By September 2016, Tunisia was the only country involved in the Arab Spring that had become a democracy.

Numerous factors lay behind the Arab Spring: dissatisfaction with governments, dictatorships, corruption, economic decline, unemployment, inequalities in wealth, food shortages and escalating food prices. Widespread access to social media networks made the Arab Spring possible in countries such as Tunisia and Egypt, whereas in Yemen and Libya people communicated through the traditional forms of media.

A number of reasons have been put forward to explain what has determined change in some areas but not others. They include:

- strong civil societies—countries with strong civil societies, such as Tunisia, were more successful because they were able to transform the country after political change

- the degree of state censorship—in countries where Al Jazeera and the BBC provided widespread coverage, such as Egypt, mass violence by the government and the military was suppressed

- social media—countries with greater access to social media were more able to mobilize support for the protests.

Places where restricted freedoms have been challenged:
The Rohingya of Myanmar

The Rohingya are one of the world's most persecuted minorities, without citizenship. In 2014, the Myanmar government expelled humanitarian groups, thus preventing health care and aid for the Rohingya.

There has been considerable political change in Myanmar, from the decades of authoritarian military rule to the democratic election of Aung San Suu Kyi of the National League for Democracy party in November 2015, and her swearing in as president in March 2016.

In the elections, over 6,000 candidates represented more than 90 political parties. However, 75 candidates were disqualified for failing to meet citizenship or residence criteria, including all of the Rohingya candidates and most Muslims.

Despite publicity, the Rohingya continue to be restricted and discriminated against.

Concept link

POWER: Global interactions can be rejected by governments and their citizens in order to preserve national culture and protect domestic industry and resources. Global recession and political conflict are two examples where power is exerted via the closing of physical borders in order to safeguard domestic interests. Alternatively, the power of public protest (such as when there is an upwelling of support for greater freedom) can be an agent of change and encourage governments to enact change.

Test yourself

5.28 Describe the variations in free access to the internet according to figure 5.3.4.   [3]

5.29 Using examples, **explain** how social media can influence social development.   [8]

QUESTION PRACTICE

a) Analyse the global and local cultural influence of diaspora populations.   [12]

b) Examine how constraints on global interactions have been challenged in one or more named countries.   [16]

How do I approach these questions?

a) The command term is analyse. Therefore, you should break down the question to bring out the essential elements or structure. In this case, you should show how diaspora populations have influenced culture on a global and local scale. To reach the top level, you should have examples to back up the points you make.

The focus should be on a diaspora population: which countries have they gone to, and what aspects of their culture have they brought with them.

b) The command term is "examine": consider an argument or a concept in a way that uncovers the assumptions and interrelationships of the issue. In this question you may examine the constraints that are imposed on people, and how these constraints are being challenged, in one or more named countries.

Good answers are likely to show the nature of the constraints and how they are being challenged. They may show how the constraints and the challenges have changed over time. They may refer to the unequal power that different players have.

▲ Introduction—good detail. Focuses on global scale

▼ Government focus rather than diaspora

▲ But attempting to support diaspora population and raise awareness of Korean culture. Helps explain the influence on a global scale

▲ Evidence of Korean culture spreading, and detail of its success/appeal via YouTube. Good coverage of the global scale

▼ A bit more needed on Korean culture apart from Psy's Gangnam Style music —e.g. food, literature, art

▲ General point about different cultures being accepted in a foreign country

▲ Good local-scale example

▲ Located example—good detail

▲ Good description of cultural facilities for Koreans and others

▲ Valid explanation for the location of the Korean community in this location, and its growth

▲ Good conclusion—refers to both global and local scale

a) There are over seven million Koreans living outside of Korea. About 80% of these are found in just three countries - China (2.5 million), USA (2.4 million) and Japan (0.45 million). The Korean Culture and Information Service (KOCIS) runs 32 cultural centres in 27 countries. Its main aim is to enhance the image of Korea's "national brand" by promoting Korean heritage and arts through these cultural centres. The spread of Hallyu (Korean wave) has been accelerated through K-pop (Korean pop music) through artists such as Psy's Gangnam Style music video (the first to reach one billion views on YouTube), and Girls Generation SNSD.

Each culture has different practices, beliefs, values and traditions upheld to a greater or lesser extent by members of that group. Many western governments encourage the idea of "integration". This refers to incorporating minorities and their cultures into that of the nation. The nation is seen as having its own culture that minorities must accept and live by, whilst retaining their own ethnic and religious cultures.

There are also local impacts. New Malden is a town in south-west London that has one of the most densely populated areas of Korean residents outside of South Korea. There are around 10,000 Koreans (of which about 600 are from North Korea) out of New Malden's population of 20,000. Many of those living in New Malden are working for South Korean companies. New Malden has Korean places of worship, nursery schools, about 20 Korean restaurants/cafes, shops, supermarket and a karaoke bar. The reason for the concentration of Koreans here is part historic and part accessibility. New Malden has a good rail connection with central London. It was also the original site for the Embassy of South Korea and Samsung Electronics had its UK headquarters there. Thus, Korean culture can be seen to have an influence at a global scale, and at a local scale, even in a country that does not have a particularly large number of Korean diaspora.

Very good answer – better on a local than a global scale. Good place detail. Explains the reasons for their location in New Malden.

Marks 10/12

▲ Introduction identifies four themes

b) There are many constraints on global interactions such as the rise in anti-immigration groups, protectionism and national trade restrictions. These are happening in many parts of the world, including HICs and LICs, and are having a major impact on global interactions.

There are many reasons for the rise of anti-immigration groups including the perceived threats over competition for jobs, and the cost of housing, education and health care. In some cases, notably in LICs and MICs, environmental issues may also be a concern, as a result of rapid population growth. Some argue that certain immigrant groups isolate themselves from society and refuse to integrate into mainstream society.

In 2017, US President Donald Trump prevented migrants from seven, mainly Muslim, countries from entering the USA, although his decision was overturned by a number of US courts. He reiterated his intention of having a wall built between the USA and Mexico to reduce migration. Part of the reason for the UK voting to leave the European Union was due to anti-immigration sympathy amongst many British people.

> ▲ One constraint identified and developed. Good detail

> ▼ However, nothing about how the constraint has been challenged

Trade restrictions are a form of protectionism: most trade restrictions place an additional charge on traded goods to make home goods more competitive. Most economists would argue that trade restrictions increase inefficiency and lead to less choice for consumers, although they may help a country to industrialize. Trade barriers have been criticized as well, as they often affect LICs. Protectionism reduces trade between countries. Since 2008, 70% of the 20 OECD nations have imposed restrictive trade policies in response to the global economic slowdown.

For example, in 2015, the USA imposed a 256% tariff on Chinese steel and a 522% tariff on cold-rolled steel. Although a meeting of the G7 in June 2018 issued a pledge to fight protectionism and cut trade barriers, it was immediately followed by the imposition of tariffs against the USA by Canada, in response to US tariffs against Canada's steel and aluminium industries. Organizations such as the IMF and the World Trade Organisation try to mediate over trade restrictions but they move very slowly and often it takes years for any response.

> ▲ Second constraint identified and developed. Again, good detail present

> ▼ Some reference to "challenge", but lacks detail

The Arab Spring refers to the range of demonstrations, protests, riots and civil wars that spread through countries in the Middle East and North Africa after 2010. Most of the revolutions and protests were over by 2012, although the ongoing conflict in Syria is an important exception. By September 2016, the only country involved in the Arab Spring to become a democracy

▲ Identifies a challenge to the constraints in the Middle East and North Africa with good detail

was Tunisia. Numerous factors lay behind the Arab Spring: dissatisfaction with governments, dictatorships, corruption, economic decline, unemployment, inequalities in wealth, food shortages and escalating food prices. Widespread access to social media networks made the Arab Spring possible in countries such as Tunisia and Egypt, whereas in Yemen and Libya, people communicated through the traditional forms of media.

Following the protest and changes associated with the Arab Spring was the so-called Arab Winter, a wave of violence, instability and economic decline. The Arab Spring has thus had mixed success. For some, there has been greater freedom compared with the restrictions before, as in Tunisia and Egypt. For others, the Arab Spring has led to a collapse of law and social order, as in Syria and Libya.

▲ Some development of the challenge—where the challenge succeeded and where it failed

A number of reasons have been put forward to explain what has determined success in some areas but not others. They include:
• strong civil societies—countries with strong civil societies such as Tunisia were more successful than those without because they were able to transform the country after political change;
• the degree of state censorship—in countries where Al Jazeera and the BBC provided widespread coverage, such as Egypt, mass violence by the government and the military was suppressed, in contrast to countries such as Libya and Syria, where there was less television reporting;
• social media—countries with greater access to social media were more able to mobilize support for the protests

▲ More detailed reasons for the success/failure of the challenge of the Arab Spring

• support of the national military—in Egypt and Tunisia, the military supported the protesters in removing the government, whereas in Libya and Syria the military have contributed to civil war;

▼ However, the use of bullet points prevents detailed evaluation and synthesis

• the mobilization of the middle class—countries with a strong, vocal middle class were more likely to see political change than countries with a weak or limited middle class.

▼ Very brief conclusion and challenges have not been a major focus of this essay

Thus there are many constraints on global interactions, and there have been numerous challenges to overcome these constraints.

Good on the Arab Spring as a way in which constraints on global interactions have been challenged—but elsewhere the constraints are described, but not the challenges. For a higher mark, there should be a wider range of applied knowledge being synthesized. For example, there could be more mention of physical constraints on interactions and attempts to overcome these using mobile phones and ICT.

Marks 9/16

6 GLOBAL RISKS AND RESILIENCE (HL ONLY)

This unit examines some of the threats to individuals and businesses, and the political and economic sovereignty of states. It also analyses transboundary pollution, and the environmental impacts of global flows, such as agribusiness. It considers ways in which civil society raises awareness about economic and social risks of global interactions.

You should be able to show:

✔ how technological and globalizing **processes** create new geopolitical and economic risks for individuals and societies;

✔ how global interactions create environmental risks for particular **places** and people;

✔ new and emerging **possibilities** for managing global risks.

6.1 GEOPOLITICAL AND ECONOMIC RISKS

You should be able to show how technological and globalizing processes create new geopolitical and economic risks for individuals and societies:

✔ Threats to individuals and businesses:

  ✔ hacking, identity theft and the implications of surveillance for personal freedoms;

  ✔ political, economic and physical risks to global supply chain flows;

✔ New and emerging threats to the political and economic sovereignty of states:

  ✔ profit repatriation and tax avoidance by TNCs and wealthy individuals;

  ✔ disruptive technological innovations, such as drones and 3D printing;

✔ The correlation between increased globalization and renewed nationalism/tribalization;

  ✔ Two detailed examples to illustrate geopolitical tension/conflict.

> • **Cybercrime** – criminal activity using the internet/computers/computing.
>
> • **Drone** – unmanned aerial vehicle (UAV).
>
> • **Profit repatriation** – the return of a company's foreign-earned profits or financial assets to that company's home country.
>
> • **3D printing (or additive manufacturing)** – the creation of a physical object from a digital model by laying down a sequence of layers.

Threats to individuals and businesses

Hacking

A major hacking "event" occurred in the UK in May 2017, when a number of National Health Service (NHS) trusts were left vulnerable in a ransomware attack. More than one-third of trusts in the UK were disrupted, resulted in nearly 7,000 appointments being cancelled. The malware, named WannaCry, spread to more than 150 countries in a worldwide ransomware outbreak.

Test yourself

Study figure 6.1.1.

6.1 Describe the trend in identity theft in the USA between 2007 and 2016.   [2]

6.2 Determine the number of victims of identity theft and the cost/time to resolve identity theft issues.   [1]

As digital technologies are increasingly embedded in everyday objects (the so-called Internet of Things) there is increased potential for hackers to monitor and affect operations. This may mean that increased surveillance is required to maintain security.

Identity theft

Identity crime is a generic term to describe a range of crimes from complete life theft to credit card theft and subsequent fraud. Identity theft is the acquisition of identity-related data through "phishing", data breach/theft, deception and accidental loss. Technology not only creates criminal opportunities, it also creates ways of detecting them. Contactless payment schemes have increased the potential for identity crime, as payment can be made with a simple tap of the card.

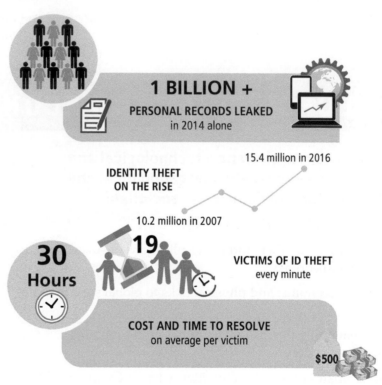

1 BILLION +
PERSONAL RECORDS LEAKED
in 2014 alone

IDENTITY THEFT ON THE RISE

15.4 million in 2016

10.2 million in 2007

30 Hours

19

VICTIMS OF ID THEFT
every minute

COST AND TIME TO RESOLVE
on average per victim

$500

▲ **Figure 6.1.1.** Identity theft in the USA

Implications of surveillance for personal freedoms

Mass surveillance is the practice of spying on an entire, or a significant part of a, population. It can range from CCTV and email interceptions to wire-tapping and computer hacking. The Council of Europe stated that mass surveillance practices are a fundamental threat to human rights and violate the right to privacy enshrined in the European Convention on Human Rights. It also suggests that the British laws enabling the wide-ranging powers of GCHQ (Government Communications Headquarters) are incompatible with the European Convention on Human Rights.

The internet and social media allow people to communicate more freely than ever before. But people are under more surveillance than ever before from governments and commercial organizations. In 2013, Edward Snowden, a former US intelligence analyst, leaked top secret files which exposed the existence of massive surveillance by the USA and UK. The debate concerning how far the government should interfere with human rights for the sake of national security continues.

Political, economic and physical risks to global supply chain flows

Economic risks	Explanation
Asset price rises in a major economy	Unsustainably overpriced assets such as commodities, housing and shares in a major economy or region
Failure/shortfall of critical infrastructure	Failure to adequately invest in, upgrade and/or secure infrastructure networks (for example, energy, transportation and communications)
High structural unemployment or underemployment	A sustained high level of unemployment or underutilization of the productive capacity of the employed population
Illicit trade (for example, illicit financial flows, tax evasion, human trafficking and organized crime)	Large-scale activities outside the legal framework that undermine social interactions, regional or international collaboration, and global growth
Severe energy price shock (increase or decrease)	Significant energy price increases or decreases that place further economic pressures on highly energy-dependent industries and consumers
Environmental risks	
Extreme weather events (for example, floods and storms)	Cause major property, infrastructure and/or environmental damage as well as loss of human life
Major biodiversity loss and ecosystem collapse (terrestrial or marine)	Result in irreversible consequences for the environment and severely depleted resources for humanity as well as industries
Major natural disasters (for example, earthquake, tsunami, volcanic eruption, geomagnetic storms)	Cause major property, infrastructure and/or environmental damage as well as loss of human life
Human-made environmental damage and disasters (for example, oil spills and radioactive contamination)	Failure to prevent major human-made damage and disasters, including environmental crime, causing harm to human lives and health, infrastructure, property, economic activity and the environment
Political risks	
Failure of national governance	Inability to govern a nation of geopolitical importance as a result of weak rule of law, corruption or political deadlock
Failure of regional or global governance	Inability of regional or global institutions to resolve issues of economic, geopolitical or environmental importance
Interstate conflict with regional consequences	A bilateral or multilateral dispute between states that escalates into economic (for example, trade/currency wars, resource nationalization), military, cyber, societal or other conflict
Large-scale terrorist attacks	Individuals or non-state groups with political or religious goals that successfully inflict large-scale human or material damage
Weapons of mass destruction	The deployment of nuclear, chemical, biological and radiological technologies and materials, creating international crises and the potential for significant destruction of property and infrastructure

▲ **Table 6.1.1.** Political, economic and physical risks to global supply chain flows

>> **Assessment tip**

Try to add examples to each of the risks shown in table 6.1.1 (start by thinking of your home country), so that you can use them as "detailed locational support" in an essay on risks to global supply chain flows.

Test yourself

6.3 Compare the advantages and disadvantages of global supply chains.   [3+3]

6.4 Explain how the main environmental risks can affect global supply chains.   [2+2+2]

New and emerging threats to the political and economic sovereignty of states

Profit repatriation and tax avoidance by TNCs and wealthy individuals

According to the US Public Interest Research Group (PIRG) the USA's largest companies hold nearly US$2.5 trillion of accumulated profits in offshore accounts to avoid paying tax. The money is held in over 10,000 tax havens including Bermuda, Ireland, Luxembourg and the Netherlands.

PIRG estimates that the US government is missing out on US$100 million a year in taxes from nearly 400 Fortune 500 companies. It has calculated that, if these companies paid back all of the tax they would have had to pay if they did not hold their profits overseas, it would amount to US$717 billion, which is equivalent to the GDP of the 19th largest country (larger than Switzerland or Saudi Arabia).

In 2017, a set of leaked documents known as the "Paradise Papers" revealed the tax affairs of the wealthy and TNCs, and their complex structures to avoid paying tax. Such structures, as used by Apple, Facebook and Google, are legal. Companies have a duty to their shareholders to operate efficiently and maximize profits. Many famous musicians and actors are also paid through offshore trusts. Although it may be legal, it is unethical. Tax avoidance is not victimless. It results in governments having less finance to support schools, health care, social welfare and infrastructural developments.

Many wealthy individuals donate money to charities or set up their own charities. Some of this may be very genuine, but it can also be a way of avoiding paying tax.

Disruptive technological innovations: Drones

Drones (unmanned aerial vehicles, or UAVs for short) are generally used for the three Ds—dull, dirty and dangerous work. They have a number of disadvantages. For example, drones can cause an invasion of privacy. In military situations, operators are disconnected from the war zone where the drones are being used, and therefore can be desensitized to the act of killing. Drones can also cause civilian deaths in war. Across the world, countries need to be aware of the threat from drones being used by other nations and terrorist groups on them.

▲ **Figure 6.1.2.** Drones could have a positive impact on farming

However, one advantage is that drones can enter environments dangerous to humans. They can fly for long distances in inhospitable environments. They do not require a qualified pilot on board. Drones can be used for commercial purposes, for example, farmers can use them for precision agriculture; they can also deliver medicines to remote areas, monitor environmental change, track criminals, monitor chemical hazards and detect explosives. Compared with aircraft, drone operation is relatively low cost and low risk. In the future, it is likely that drones will become much more commonplace in delivering consumer items and also medication, especially to remote rural areas.

Disruptive technological innovations: 3D printing

3D printing (or additive manufacturing) enables producers to manufacture a range of products using a 3D printer and raw materials such as plastics, wood, paper, resin and glass. 3D technology can manufacture goods quickly, which reduces the time needed for designs to be tested. However, the set-up costs are high. It reduces the need for storage of large amounts of raw materials, and creates many opportunities for highly skilled designers who are familiar with the technology. One of the main breakthroughs in 3D printing technology is the manufacture of human organs (bioprinting).

Eindhoven in the Netherlands is the world's first city to have habitable homes made with the use of 3D printers. Project Milestone was developed, in part, due to the lack of skilled bricklayers. The process has cut costs and environmental damage. The 3D printer that was used is essentially a huge robotic arm with a nozzle that ejects a specially formulated cement.

However, 3D printers use up a large amount of energy, so they are better suited for small production runs. 3D printing is also expensive. It mainly uses plastic, and the printers may generate toxic emissions. They may also emit volatile organic compounds during printing, and these have been linked to respiratory illnesses, heart disease and cancer. 3D printers are slow, especially when a variety of raw materials are used. 3D printers may also be used for criminal purposes: they may be used to produce weapons and to create card readers for bank machines.

The correlation between increased globalization and renewed nationalism

The opposite of globalization is nationalism, and it is becoming widespread in North America and Europe, which previously had been considered among the main winners from globalization. However, people from around the world are now looking to their nation state to protect them from global competitors, which are often TNCs rather than other countries.

Geopolitical tension/conflict: Rising nationalism in Europe

Nationalism is making a reappearance. Contemporary Europe is, fundamentally, a peaceful and prosperous continent, and the EU provides a framework for extremely close cooperation among national governments. Nevertheless, nationalistic feelings are growing, as seen by governments wishing to defend their national self-interest within the EU (for example, the United Kingdom's referendum to leave the EU), and the rise of right-wing populist nationalist political groups.

Radical right-wing populism in the EU is largely based on Islamophobia and anti-immigration. It draws on angry attitudes among sections of society that struggle with multiculturalism or think that they are losing out in a globalized economy. A common criticism is that the economic policies of these right-wings groups do not extend beyond a rage at the euro, free trade and foreign populations.

Geopolitical tension/conflict: Economic nationalism in USA and China

Economic theory suggests that free trade is the best policy for all countries (though there are some who do not agree with this view) and trade wars are mutually destructive and have a wider known knock-on effect.

In 2018, US President Trump announced plans to impose 25% tariffs on US$50 billion worth of Chinese imports. China responded by imposing the same level of tariffs on a similar amount of US imports. President Trump then announced further tariffs on US$400 billion worth of imports. Currently, China imports about US$130 billion worth of US goods, compared with the USA's import of over US$500 billion worth of goods from China.

Chinese investment in the USA is around US$165 billion, compared with US investment in China of over US$625 billion. Thus, the USA is vulnerable to the Chinese imposing restrictions on US investment in China. Moreover, the impacts would affect more than just the USA and China. Any country that supplies goods to the supply chain in China or the USA would be affected, as would any country that imports goods from either country. Global inflation would likely result.

> **Test yourself**
>
> **6.5** Using examples, **analyse** the problems associated with tax avoidance by TNCs and wealthy individuals. [6]
>
> **6.6** Using an example, **explain** the rise of nationalism. [6]

6.2 ENVIRONMENTAL RISKS

- **Transboundary pollution (TBP)** – pollution that originates in one country but affects another country.

- **Carbon footprint** – a measure of how much carbon is used to produce, store, transport and sell goods to consumers.

- **Agribusiness** – large-scale, commercial, intensive farming.

You should be able to show how global interactions create environmental risks for particular places and people:

✔ Transboundary pollution (TBP) affecting a large area/more than one country;

  ✔ TBP case study including the consequences and possible responses;

✔ Environmental impacts of global flows at varying scales:

  ✔ localized pollution, including impacts along shipping lanes;

  ✔ carbon footprints for global flows of food, goods and people;

✔ Environmental issues linked with the global shift of industry:

  ✔ polluting manufacturing industries;

  ✔ food production systems for global agribusiness.

Transboundary pollution

Case study: Chernobyl disaster, April 1986

>> **Assessment tip**

Although it is recommended that your case studies come from within your own lifetime, sometimes the best examples are a little older. Although there have been examples of transboundary pollution that have occurred in more recent years, this example has a very clear spatial variation in its impact, and it is still having an impact on those who live in the area.

The Chernobyl disaster was the explosion of a nuclear reactor in Ukraine in 1986. A combination of design flaws and human error contributed to the Chernobyl disaster. First, there were design drawbacks with the reactor. Second, human error due to poor supervision led to unstable operations.

Radionuclides, normally contained within the reactor core, were released into the atmosphere for nearly ten days. Widespread distribution of airborne radionuclides resulted across most of western Europe. This led to the contamination of soil, plants and animals, and the contamination of foodstuffs.

Firefighters fought the fire at very close quarters. All of the firefighters received serious radiation doses. There were 31 deaths and over 200 cases of radiation burns. The health effects of the disaster are still emerging; for example, there has been a sharp increase in the number of throat cancers found in those who lived in the most contaminated areas at the time of the accident.

In order to isolate the exposed reactor, a "sarcophagus" (coffin) was built to contain it. However, due to the difficult working conditions, there are gaps in the container, although these are regularly monitored for radiation.

Everybody from within 30 km of the reactor was evacuated. After the evacuation, decontamination work began. All soil to a depth of 15 cm was removed, and all buildings had to be cleaned. These measures were later found to be of limited effectiveness. To prevent the contaminated land being washed away, and radionuclides seeping into rivers and the Kiev reservoir, 140 dykes and dams were built.

Environmental impacts of global flows at varying scales

Localized pollution, including impacts along shipping lanes

Shipping is a growing sector but one of the least regulated sources of atmospheric pollutants. Shipping makes a significant contribution to emissions of sulfur dioxide (SO_2) and oxides of nitrogen (NO_x), and to $PM_{2.5}$ and PM_{10} (particulate matter with a diameter of less than 2.5 and 10 micrometres respectively).

Test yourself

6.7 Describe how the hazards associated with the Chernobyl explosion varied spatially. [4]

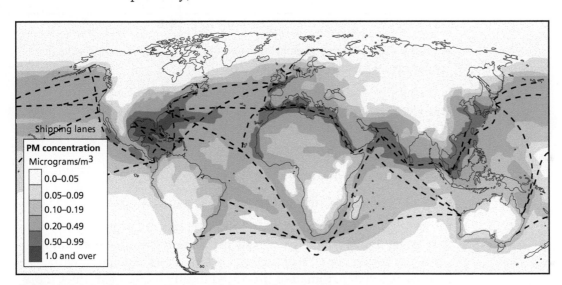

▲ **Figure 6.2.1.** Shipping pollution and the world's main shipping routes

▼ **Table 6.2.1.** Proportion of annual emissions produced from shipping

Area	Year	NO_x	SO_2	$PM_{2.5}$	PM_{10}
Europe	2009	23%	16%	7.9%	5.5%
UK	2011	45%	40%	21%	15%
UK	2020	73%	14%	21%	15%

Source of data: Department for Environment, Food and Rural Affairs, UK

Shipping uses diesel engines almost exclusively. The International Maritime Organization has set a global cap of 4.5% for the sulfur content of marine fuel oil. There are clear benefits from the reduction of sulfur emissions.

Emissions of black carbon from ships is a big issue. On a global scale, shipping emissions only account for 1–2% of total emissions. However, unlike the well-mixed greenhouse gases, the climate impacts of short-lived black carbon are regional, with larger impacts nearer the areas of higher emissions. This is of particular concern in the Arctic and other areas of the cryosphere (ice and snow environments) where the deposition of black carbon on snow and ice can reduce the albedo and add to the direct warming effect of black carbon.

Carbon footprint of the global flows of goods and services

According to the Carbon Trust, about 25% of greenhouse gases are embodied in goods and services which "flow" between the country of production and the country of consumption via international trade.

HICs are typically net importers of embodied carbon emissions, whereas LICs and emerging economies are generally net exporters of CO_2 emissions.

Transport is the largest end-use contributor towards global warming in the USA. CO_2 emissions from transport exceeded 2 billion tonnes as far back as 2007. Transport has a significant impact in the food and drink sector because food is often transported long distances and by air. Even produce grown and consumed within North America travels on average 2,000 km from source to point of sale.

Content link
Relate to this information to the discussion of the water–food–energy nexus in unit 3.2.

	Megajoules per tonne-km	kg CO_2e per tonne-km
International shipping	0.2	0.14
Inland water	0.3	0.21
Rail	0.3	0.18
Truck	2.7	1.8
Air	10	6.8

▲ **Table 6.2.2.** Energy and emissions per tonne-km (CO_2e is carbon emissions converted to the equivalent CO_2 emissions)

Food supply chains are often long and complex. Cultivation may be restricted spatially (by area) and by time (seasonally). There are three options for providing fresh produce when it is out of season locally:

*   sourcing from distant growing areas
*   using long-term storage
*   cultivating in a protected area such as a greenhouse.

Importing produce may result in lower overall emissions than harvesting and storing local products for many months. For example, storage accounts for 60% of the carbon emissions associated with carrots; in Sweden, tomatoes produced locally require ten times more energy than field-grown tomatoes imported from southern Europe. Highly perishable food often requires cooling, refrigeration or freezing during transport and/or storage.

For food products derived from animals, such as meat and dairy goods, transportation contributes only a small proportion of their carbon emissions, but for plant-based products, such as fruit and vegetables, transport contributes a higher proportion of their carbon emissions.

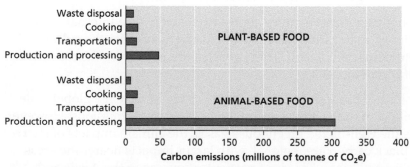

▲ **Figure 6.2.2.** Life cycle carbon emissions (millions of tonnes of CO_2e) for plant-based foods (top) and animal-based foods (bottom) in the US

Source of data: Adapted from *Green Technologies in Food Production and Processing* (2012)

Test yourself

6.8 Explain why shipping is associated with high rates of pollution. [6]

6.9 Suggest why importing food may have a lower environmental cost than locally produced food. [6]

Carbon footprints of food are likely to increase due to population growth, rising incomes and increasing levels of urbanization.

The carbon footprint of tourism is about four times larger than previously thought. Researchers from the University of Sydney, University of Queensland and National Cheng Kung University examined the entire supply chain of tourism, including transportation, accommodation, food and beverages, souvenirs, clothing, cosmetics and other goods, from 2009 to 2013. They found that global tourism produces about 8% of global greenhouse gas emissions. They identified carbon flows between 160 countries from 2009 to 2013, and found that tourism-related emissions increased by around 15% over that period, from 3.9 gigatonnes (Gt) of carbon-dioxide equivalent (CO_2e) to 4.5 Gt. They also suggested that carbon emissions from global tourism would increase to about 6.5 Gt by 2025 due to rising incomes.

Environmental issues linked with the global shift of industry

Polluting manufacturing industries

Since the 1970s, many emerging economies have adopted lax environmental standards to attract foreign firms to move production there. However, these policies may attract large-scale polluting industries to the host countries.

A significant number of US firms have reduced their pollution at home by offshoring production to less developed and less regulated countries. For example, between 17% and 36% of four major air pollutants emitted in China comes from export-orientated production. Of these emissions, over 20% come from the production of goods for the USA.

Of all the goods imported by US manufacturing firms, the proportion produced in emerging economies rose from 7% in 1992 to 35% in 2016. At the same time, toxic air emissions from manufacturing industries in the USA decreased by over 50%.

US companies that contribute to offshore pollution are not violating environmental laws either at home or in their host countries. They are taking advantage of those nations' lower environmental and labour standards and letting the host countries bear the costs.

It is not just HICs that relocate polluting industries overseas. Some emerging economies have started to do so, too. The Chinese Hebei Iron and Steel Company has announced plans to relocate some of the province's steel, cement and glass production to Africa, Latin America, Eastern Europe and other parts of Asia. Capacity for 20 million tonnes of steel and 30 million tonnes of cement will be moved overseas by 2023. It is building a plant capable of making 5 million tonnes annually in South Africa, and is likely to shut mills in Hebei, currently home to seven of China's most polluted cities. Just as rich countries outsourced their pollution to China, mainly light manufacturing, China has got to the point in development where it wants to start exporting pollution too, by building steel and other factories in poorer countries.

Food production systems for global agribusiness

Agribusiness results in the conversion and degradation of natural ecosystems around the world to highly simplified agricultural ecosystems. Agribusiness is often based on monoculture (the growing of single crop varieties over large areas).

Concept link

PLACES: Environmental incidents, such as the Chernobyl nuclear disaster in 1986, demonstrate how events that happen in one country create transboundary pollution incidents, which then have negative impacts on places for a considerable amount of time.

The flow of goods between places will create pollution at the production source, when items are produced using non-renewable energy sources and using scarce water supplies. The movement of products along shipping routes using some of the world's largest ships will produce more pollution than all of the vehicles in the world combined, thus affecting places at a local scale. Places gain or lose industry due to market forces and this can result in benefits, such as a reduction in air pollution as manufacturing industry relocates elsewhere.

Pesticides are any substance used to destroy unwanted organisms. However, they have environmental and health impacts. They may be toxic to non-target organisms. They can increase surface and groundwater pollution.

The use of nitrogen (N), phosphorus (P) and potassium (K) fertilizers (sometimes referred to as NPK fertilizers) is a major source of pollution of water. It can lead to eutrophication and the creation of "dead zones" in coastal and inland waters. The manufacture of fertilizers is energy intensive and requires a substantial input of fossil fuels.

The number and size of concentrated animal feeding operations (CAFOs or feedlots) has increased. These produce large amounts of manure and significant nitrate pollution of air, water and groundwater.

There are also significant demands on water, leading to groundwater depletion, salinization and decreases in water availability for other users. In western USA, some 90% of water is used for agriculture.

Test yourself

6.10 Explain the global shift of polluting manufacturing industries. [6]

6.3 LOCAL AND GLOBAL RESILIENCE

- **Crowd-sourcing** – the process of sourcing ideas, services, funding or content from the public in order to maximize the benefit of a large group's collective assets.

- **Cyber security** – the protection of information systems, hardware and software from theft or damage.

- **E-passport** – electronic passport containing a computer chip with details about the owner.

- **Reshoring** – the relocation to the home country of a business's or company's operations that were overseas.

- **Resilience** – the ability to protect lives, livelihoods and infrastructure from destruction, and the ability to recover after an event.

You should be able to show new and emerging possibilities for managing global risks:

✔ The success of international civil society organizations in attempting to raise awareness about, and find solutions for, environmental and social risks associated with global interactions;

   ✔ Detailed examples of one environmental and one social civil society organization action;

✔ Strategies to build resilience:

   ✔ reshoring of economic activity by TNCs;

   ✔ use of crowd-sourcing technologies to build resilience by government and civil society;

   ✔ new technologies for the management of global flows of data and people, including cybersecurity and e-passports.

The success of international civil society organizations in attempting to raise awareness about, and find solutions for, environmental and social risks associated with global interactions

Case study: Environmental organization action—WWF and the Great Barrier Reef

Australia's Great Barrier Reef is a World Heritage Area and one of the largest living structures in the world. It covers an area of over 344,400 km², and is one of the world's most popular tourist attractions. Tourism to the reef earns Australia some AU$6 billion annually and provides around 69,000 jobs.

However, the reef is extremely vulnerable. Since 1990, it has lost around half of its coral cover; pollution has led to deadly starfish outbreaks; global warming has led to coral bleaching; there are even plans to expand several ports along the Great Barrier Reef in order to export coal from the Galilee Basin to India. Increasing the size of ports will lead to more dredging of the reef and increased

>> **Assessment tip**

Your case studies for this section must cover environmental and social issues. You could have a single case study that covers both aspects, or you could provide two separate accounts, as here.

shipping pollution. Pollution from farms (producing crops and livestock for the global market) gets washed off the land and smothers coral and seagrass beds, reducing the clarity of the water. It also leads to population explosions of crown-of-thorns starfish, while nitrogen runoff from farms leads to algal blooms, and in some cases dead zones.

WWF Australia is committed to improving conservation methods, increasing awareness about the issues and supporting initiatives to manage fisheries and preserve key species. WWF Australia is trying to halt and reverse the decline in species and ecosystem health in the reef, and to reduce the impact of climate change on the reef. They have called for reform of Queensland's fishing management system, and have encouraged people to "adopt" a turtle.

> **Content link**
> Management of coral reefs is discussed in option B.3.

Case study: Civil society organization action—Survival International and Aboriginal people

The Aboriginal population has inhabited Australia for around 45,000 years. However, since the 1800s their population numbers have crashed, and the quality of their lives has been much reduced. It is thought that there were around one million Aboriginal people in Australia when the British first arrived in the country. As a result of the diseases introduced and massacres of Aboriginal people by the colonists, that population fell to just 60,000 between 1850 and 1951.

Before the British occupied Australia, the Aboriginal people were a race of hunter-gatherers and farmers, and lived in the rural parts of the country. As a result of losing their traditional lands to the British, many of them have had no choice but to settle in the poorest urban areas. For some, this has had a terrible impact in a number of ways. The traditional hunting/farming skills Aboriginal people were brought up with would have no place in an urban environment; a lack of education and local employment opportunities means that options are limited for young people and adults.

Furthermore, unsanitary living conditions have resulted in high infant mortality rates and lower life expectancy. For children born in 2010–2012, life expectancy for Aboriginal men and women is projected to be 69.1 and 73.7 years respectively, compared with 79.7 and 83.1 years for non-indigenous Australians. From 2011–2015, the infant mortality rate for Aboriginal populations was 6.1 per 1000 live births, compared with 3.3 per 1000 live births for non-indigenous infants. Other social issues faced by the Aboriginal people living in the poorest parts of the country include poor diet and alcohol and drug abuse.

Survival International has provided funds for projects that enable Aboriginal people to return to rural areas (known as homelands) where they have traditional ownership or historical association, and supports them to gain native titles to these lands in the courts and in parliament. The organization successfully campaigned with the Mirarr people of the Northern Territory against a mining company that wanted to open a uranium mine on land that is sacred to the Mirarr. In addition, it raises awareness of the issues through its website and various publications.

Test yourself

6.11 Evaluate the role of one international civil society organization in attempting to raise awareness about, and find solutions for, social risks associated with global interactions. [8]

Strategies to build resilience

Reshoring of economic activity by TNCs

US manufacturing has experienced a wave of resurgence, much of it due to reshoring. Companies that have returned their production to the USA include General Electric and Ford. A number of factors are driving the reshoring, including rising wages in China, currency fluctuations

193

and the impact of higher energy costs on shipping. Disadvantages of offshoring include the geographic distances between places of research or engineering and the location of production, supply chain disruptions and concerns over intellectual property theft by the outsourced manufacturers.

Reshored items commonly include heavy machinery (which is costly to ship), expensive items subject to changes in consumer demand (for example, clothing and home furnishings), and products with which there are potential contamination issues during the production or shipping process (for example, formula milk and food items).

In Europe, after concerns regarding higher tariffs as a result of Brexit, almost 50% of businesses have started looking to replace British suppliers with competitors from within the EU. More than 25% of European supply chain managers intend to reshore all or part of their supply chains to Europe, with 46% anticipating an increased proportion of their supply chain being removed from the UK. The effects of the devaluation of the UK currency have made overseas supply chains for British companies more expensive.

At the same time, around 33% of British firms that use EU-based suppliers claim to be looking for UK replacements. This may benefit British manufacturing. The car industry employs some 170,000 in Britain, and companies such as Honda, Nissan and Jaguar Land Rover have complex supply chains and engineering firms all over the EU. They want to buy more goods locally—Nissan wants to double the share (by value) of parts sourced in Britain from 40% to 80%.

Use of crowd-sourcing technologies to build resilience by government and civil society

Crowd-sourcing is the process of sourcing ideas, services, finances and information from the public via the internet in order to benefit from the collective abilities of a large group of people. It has developed because top-down approaches, whereby the government provides the solutions to problems, have proved insufficient. Crowd-sourcing empowers people and builds mutual support.

Crowd-sourcing offers many opportunities for people to interact with each other and find solutions to old and new problems. High-speed global interactions have facilitated these interactions. Indeed, the increased interdependence and complexity of global interactions necessitates the involvement of many people from different stakeholder positions, to find solutions for problems as they arise. Crowd-sourcing is a high-technology, bottom-up approach of empowering communities around the world.

In 2008, Ushahidi was created, and it remains one of the most important open-source platform providers for crowd-sourcing crisis information. The Ushahidi platform is often used for crisis response, human rights reporting and election monitoring, such as the Nepal earthquake in 2015 and the Nigerian elections in 2011. The objective is to facilitate a better understanding of the needs of people affected by natural/human-made disasters or other issues (for example, the monitoring of safety during elections in Kenya in 2008). During the first week following the 2010 Haiti earthquake, volunteers mapped some 1,500 reports based on information from Twitter, Facebook and online news, even before they began to receive text messages. However, there was criticism of crowd-sourced crisis mapping as it was conducted in Haiti. Much of it was directed at the overflow of information and lack of coordination with humanitarian agencies for immediate action.

Test yourself

6.12 Analyse the advantages of reshoring.   [6]

>> **Assessment tip**

Real-life examples always help. Many students use crowd-sourcing for raising donations for charitable events, such as running a marathon or undertaking an arduous challenge. These are good, small-scale (local) examples which contrast well with some larger-scale events/examples such as Ushahidi.

New technologies for the management of global flows of data and people

Cybersecurity

Cybersecurity, or computer security, is the protection of information systems, hardware and software from theft or damage, as well as the protection of information on computers and related technology. The need for cybersecurity is increasing as more and more people and organizations rely on computers and the internet.

There are many threats to computer security. One of the most common is "phishing"—the attempt to obtain personal or sensitive information such as user names, passwords, bank account details and credit card details. Most computers have some level of protection, but threats are becoming more sophisticated and protection systems need to keep ahead of the threats. Common targets are large organizations, government departments, military computer systems and airline carriers. The most common prevention systems are firewalls, which stop access to internal network systems and filter out different kinds of attack.

One of the main issues regarding cybersecurity is that there is a lack of international regulations or common rules to abide by. Moreover, national security may be vulnerable to attacks from another country, making any international treaty difficult to regulate and enforce.

E-passports

An e-passport has a computer chip in it, which contains data about the owner. The advantages of the e-passport include faster checking in at airports and border clearance. E-passports may also help in crime detection as some contain biometrics such as fingerprints, which may be left at the site of a crime. The personal bioinformation on the passport will be determined by the country in which it is issued. E-passports are difficult to reproduce or forge, so security is improved. They also make it more difficult for one person to have several passports (unless they have dual citizenship).

However, there are some disadvantages associated with e-passports. Since the person who owns the passport does not have actual access to the data, if the passport is stolen, the data could be accessed and used illegally. It would be possible for someone to hack into the system and change the data.

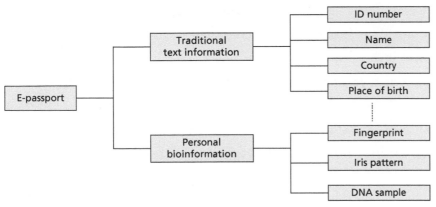

▲ **Figure 6.3.1.** Example of e-passport structure

Concept link

POSSIBILITIES: Organizations and individuals can support movements and policies in order to counter the risks presented from global interactions. International environmental organizations, such as Greenpeace, have local, national and international campaigns that are designed to reduce the negative possibilities, such as deforestation in Indonesia's rainforest. As new technology becomes more widespread, the technological risk increases. However, there is the possibility of safeguarding public and private interests using cybersecurity or facial recognition software at ports of entry, to prevent hacking and terrorism. Another possibility for facial recognition software is its use in online banking to prevent fraud and identity theft.

Test yourself

6.13 Using examples, **analyse** the advantages and disadvantages of **(a)** cybersecurity and **(b)** e-passports.   [6+6]

a) **Analyse** the carbon footprint of different global flows. [12]

b) **Examine** reasons why the global production networks of some TNCs have changed over time. [16]

How do I approach these questions?

a) The command word is "analyse", which means break down the subject of the question in order to bring out the essential elements. You will need to describe the carbon footprint of a number of flows. You may provide some comparisons in terms of size of the footprint; the direction of flows between different countries/regions; you may compare changes in carbon flows over time.

b) "Examine" means consider an argument or a concept in a way that uncovers the assumptions or interrelationships of the issue. You will need to describe a range of risks to global supply chain flow. You may suggest reasons why global supply chain flows develop (cheap products, year-round supply, standardized products); you may examine trends in risks, such as environmental degradation, or the rise of protectionism and trade barriers.

SAMPLE STUDENT ANSWER

▲ Good definition

▲ Sound scene setting – defines carbon footprint, and describes the process of getting something from source to consumption

▲ Good point about carbon footprint and importance of trade for footprint

▲ Good opening point about the contrast of HICs and LICs

▲ Good supporting data

▲ Good supporting detail – shows relative importance of trading nations/destinations

▲ Identifies the food sector and provides supporting material

▲ Good contrast between food imports and local harvest/storage

a) The carbon footprint is a measure of how much carbon is used to produce, store, transport and sell goods to consumers. It involves many stages, such as growing food, transporting it to manufacturers, processing it into a finished product, getting it to a supermarket, and the consumer making a journey to buy the goods.

About 25% of greenhouse gases are embodied in goods and services which "flow" between the country of production and the country of consumption via international trade. HICs are typically net importers of embodied carbon emissions whereas LICs and emerging economies are generally net exporters of CO_2 emissions.

The largest inter-regional flows of embodied carbon emissions are between China and North America (9%), China and Europe (8%) and China and the Rest of Asia (7%). The main regions/countries involved in footprints associated with global flows include China, Russia, Europe and North America. Some of these are source areas, e.g. China, whereas others are mainly destinations, eg North America, Europe and Japan.

In terms of food footprints, transport may account for 50% of the total carbon emissions for many fruit and vegetables but less than 10% for red meat products. Transport is not the only factor contributing to the carbon footprint of the flow of food. In fact, importing produce may result in lower overall emissions than harvesting and storing local products for many months.

Indeed, storage accounts for 60% of the carbon emissions associated with carrots. Highly perishable food often requires cooling, refrigeration or freezing during transport and/or storage. Food footprints are likely to rise due to increases in the world's population, the growing number of middle-class people and changes in diet (eg increasing proportions of meat and dairy products).

One footprint that has increased dramatically in recent years is that of tourism. Global tourism produces about 8% of global greenhouse gas emissions. The carbon footprint from tourism-related emissions increased by around 15% between 2009 and 2013, from 3.9 gigatonnes (Gt) of carbon-dioxide equivalent (CO_2e) to 4.5Gt. This rise primarily came from tourist spending on transport, shopping and food. Carbon emissions from global tourism are expected to increase to about 6.5Gt by 2025, due to rising incomes.

Thus, there are a number of global flows of many goods and services. Many of these flows link HICs with MICs and LICs. The carbon footprint of many goods and services is increasing as more people become wealthier, diets change and global tourism increases.

▼ Less here about flows

▲ But valid point that other processes may be more important than flows. Good to have material on reasons for growth. Some prediction of change given

▲ Identifies second economic sector for carbon footprints

▲ Good supporting details

▲ Good details on tourism

▼ Could say more about the reasons for its growth

Overall, a number of flows have been covered. Data on carbon footprints. Some reasons for growth. Fluent account. The factors accounting for the growth in carbon footprint (for example, population growth, middle class, changes in diet) are touched on for food, but not for tourism, hence full marks are not awarded.

Marks 9/12

b) There are many reasons why the global production networks of some TNCs have changed over time. These include economic, political, social, environmental and technological factors, such as changes in demand and supply, and the profitability of the operation.

Economic factors include currency fluctuations and trade restrictions imposed by governments, e.g. between the EU and the USA, and between the USA and China (both imposed in 2018). Costs of imports into the UK increased following the announcement of Brexit (Britain voted to leave the EU in 2016). Rising interest rates can cause havoc for TNCs that require goods at cheap prices. Debt crises in countries such as Greece, Portugal and Spain created much uncertainty there.

▲ Introduces different types of factors affecting global production networks

▲ Examples of these factors given

▲ Standard introduction which gives a hint to how the essay will be structured

▲ Contemporary support

▼ The student should really make the link between politics and economics explicit—at the moment it is just implied

▲ Economic factor 1

▲ Economic factor 2, with locational support

▲ Economic factor 3

▲ Economic factor 4

▲ Economic factor 5

Countries with strong trade unions and a history of strikes may threaten global supply-chain flows. High rates of unemployment/underemployment or more general recession may reduce demand for consumer goods. Shortages of raw materials, such as oil, may increase the cost of goods as transport costs rise.

This paragraph has a lot of good points—it would be better if it developed some of these points in more detail, perhaps with an example or a development of how it affected global production networks.

▲ Overview of political factors

▲ Political factor 1 with named support

▲ Political factor 2 with named support

▲ Political factor 3 with named support

Political factors include protectionism, trade restrictions and conflict, which can severely restrict global production networks. Supplies of goods from the Middle East and North Africa were severely affected by the so-called Arab Spring. Russia's annexation of Crimea from Ukraine led to sanctions against Russia by some western countries. Similarly, North Korea has experienced sanctions from western countries, which reduces the flow of goods into and out of North Korea.

This paragraph has fewer points but reads better as it is better supported. It would be good to bring in some details from events over the last few years (during the student's time studying IB Geography).

▲ Technological factor with located support

▲ Good example

▲ Another technological factor given

▲ Relevant examples of a technological change over time

Technological factors could affect global supply-chain flows. Transport infrastructure disruptions are unusual but ICT is very vulnerable to hacking. In 2017 Russia's largest oil company was attacked by ransomware, as was A.P. Moller-Maersk, the shipping company. The WannaCry ransomware showed how easy it is to launch a global attack, and how vulnerable some organizations are. However, technology can also increase global production networks. For example, the growth of shipping containers, from 4000–5000 TEUs in the 1980s to over 18,000 TEUs in 2013, has reduced the cost of transport significantly and allowed some TNCs to expand their operations. The combination of cheaper transport and emerging markets of China and the Pacific Rim have enabled some TNCs to grow in size.

There are many environmental factors that may disrupt global production networks. Earthquakes are a major threat in Japan and California, both of which contain many important industries. The tsunami that affected Fukushima-Daiichi led to a shortage of computer chips. Renesas Electronics Corp., the world's fifth largest chipmaker, supplied about 40% of the world market for specialized chips, known as automobile microcontrollers. It took about two months for it to start production again. Monsoon floods in Bangladesh frequently disrupt production networks there. The 2011 floods in Thailand led to shortages in hard drives after up to 1,000 factories were forced to close, leading to economic losses of $20 billion. In future, climate change may affect the production of goods (e.g. food and textiles) and submerge factories in low-lying areas.

Thus, managing production networks is a tricky task. Some of the threats may be accidental but other threats can be aggressive (political, terrorist), and the nature of threats changes (e.g. global climate change, "Brexit"). Global production networks show how intricate global interactions can be, and how changes in one part of the world can have effects in other parts.

▲ First environmental factor; very good supporting information—clearly shows the impact on the global production network of microchips

▲ Further examples of environmental impacts given; very good supporting details

▲ Another environmental factor with good supporting arguments

▲ Conclusion is relatively simple but brings the essay to a close

A good description of a range of factors (technological, economic, political, environmental) affecting global supply chain flow, including supporting details. Could have provided more on change or contrasts between places and over time. More synthesis could be present; the answer could relate transport development to the rise of middle class.

Marks 12/16

INTERNAL ASSESSMENT

Why fieldwork matters

Fieldwork is an essential part of learning geography and is compulsory for both HL and SL students. It is referred to as the *internal assessment* (IA), which means that it will be marked by your teacher and moderated by an external IB examiner.

Your fieldwork investigation is important because it will:

- help you to make sense of some of the more difficult aspects of the subject
- potentially provide useful case study material when answering an external exam question
- provide research skills that will be useful in higher education or employment
- provide an opportunity to be assessed in your geography skills in a non-examination setting
- contribute to your overall grade.

Essential information

- IA counts for 20% of the total marks at HL and 25% at SL.
- Group work is allowed for data collection.
- Fieldwork reports are written individually.
- Each report must be no more than 2,500 words long.
- It must be related to a topic on the syllabus.

Avoid these common errors in your IA:

▼ The report greatly exceeds the 2,500 word limit.

▼ The chosen topic has no spatial element.

▼ The chosen topic is not geographical.

▼ The chosen topic does not relate to the syllabus.

▼ The fieldwork question is too simplistic.

▼ The information is collected only from the internet.

▼ The survey area is too large and covers the whole region.

▼ The fieldwork information is insufficient to answer the fieldwork question.

▼ The analysis is purely descriptive.

Fieldwork research methods

Information must come from your own observations and measurements collected in the field. This **"primary information"** must form the basis of each investigation. Fieldwork should provide sufficient information to enable adequate interpretation and analysis. You may use data from third-party sources in your IA; this is known as **"secondary information"**. However, secondary information needs to be properly referenced and you should use it sparingly—the best IAs will only use secondary information to back up the primary research.

Planning and preparing your project

The success of your fieldwork will depend on your careful planning and preparation.

Choosing the right topic

The fieldwork topic must be related to the syllabus, and the most suitable topics are found within the geographic themes. The core and HL extension have very few topics that are suitable owing to their global scale. The investigation must be:

- focused upon a clearly defined fieldwork question
- confined to a small area and on a local scale
- spatial—you should consider the spatial interactions of the area chosen
- based on the collection of primary information in the field
- manageable in terms of the area covered, the time allowed and the 2,500 word limit
- able to fulfil the assessment criteria.

Choosing the right site

It is essential that you select the survey area in advance of the fieldwork investigation to ensure that it fulfils the following criteria:

- It is on a local scale, but the area covered is large enough for sufficient information to be collected.
- You can cover the area in the time allocated.
- All sites within the area are accessible at all times of day and during all seasons.

- The land is open to the public and research is permitted.
- You avoid environments that may put you in dangerous physical situations.

Where fieldwork is restricted to the school site, many successful investigations can be undertaken; for example, surveys of footpath erosion, microclimate, infiltration/ground compaction and waste management.

Devising the fieldwork question

The fieldwork question forms a basis to the research, which should allow for an investigative rather than descriptive approach. The question should be clearly focused, unambiguous and answerable. If the question is simplistic and the answer obvious, it is unlikely to be worthy of execution. However, research topics that have uncertain outcomes are still perfectly viable.

Carrying out the fieldwork

Collecting the right information

Fieldwork must involve the collection of primary information. Primary information may be qualitative or quantitative, or a combination of both. In the case of a traffic survey, qualitative data might include photographs, interviews with pedestrians and your own subjective assessment of perceived traffic hazards. Quantitative information might include traffic counts, traffic delay times, length of tailback, noise levels in decibels, or a survey of suspended particulate matter in the atmosphere. You may supplement your primary information with secondary or published information but it must not form the basis of your report. All secondary information must be appropriately credited using a comprehensive reference system.

Collecting and justifying your fieldwork methods

You must be aware of all the techniques involved and be able to critically evaluate each of them. Before you start collecting information and before you leave the survey site, make sure you have:

- marked on a map the sites where the information was collected
- recorded the date and time of collection
- recorded the weather conditions or any special event occurring on the day that might affect the results
- recorded the technique of handling a particular instrument, where it is placed, the time interval between readings, the advantages and disadvantages of the technique
- justified the choice of survey sites and their number/frequency/location
- justified the choice of method used for information collection
- justified the sampling technique used.

How to display your fieldwork information

When you have processed your fieldwork information into graphs, maps and tables, you should display these next to the text in your report to which they refer. These data should not be confined to the end of the report. Use the table below as a guide to display your data.

Method	Do	Don't
Maps	Include a map of the survey sites. Show your results at specific survey sites on this map. Annotate your map with brief analytical or descriptive comments to add value to them.	Include a national map; it is irrelevant. Include scruffy, poorly drawn maps in pencil (hand-drawn maps can be excellent—but ensure they are clear, accurate and drawn in pen). Download maps from the Internet without first modifying or adapting them for your purpose.
Graphs	Wherever possible, place a series of graphs on the same page for comparison. Use a variety of graphical techniques.	Use a monotonous series of pie charts to represent your data page by page.
Photos	Take photographs of the fieldwork techniques being carried out and annotate these to illustrate the methods. Make sure that each photograph shows the time it was taken, its location and its orientation.	Include photos of your friends and teacher unless they are strictly relevant to the investigation.
Sketches	Make sure that sketches are fully labelled/annotated and dated.	Include sketches unless they are relevant.
Generally	Make sure that all illustrations are properly referenced. Use a range of techniques, but make sure each is suitable. Map information wherever possible.	

Writing your report

Your report should be structured according to the assessment criteria shown below. The mark allocation and the recommended number of words are given for each criterion.

Criterion A: Fieldwork question

Marks available: 3 marks
Suggested word count: 300 words

For this criterion, you should introduce the fieldwork question and geographic context for the investigation. When writing this section, refer to the following points:

☐ State the fieldwork question and section of the Geography syllabus to which your investigation relates to.

☐ Give the theoretical background—what is the reason for conducting your study?

☐ Explain the geographic context—why did you pick the location? Reasons could relate to the climate, geographical features and socio-economic aspects of the area.

☐ It is essential that you include a map to show the area under investigation. It is recommended that you generate these yourself, by either hand drawing them or by generating them on a computer.

☐ Provide hypotheses for your study and add your predictions for the outcome. Justify your predictions.

Criterion B: Method(s) of investigation

Marks available: 3 marks
Suggested word count: 300 words

This criterion requires you to explain the methods of data collection used in your fieldwork. When writing this section, refer to the following points:

☐ Describe each of your methods clearly and in adequate detail such that the methods can be replicated by other students.

☐ Give justifications for the methods you have used: explain why you chose the sampling techniques you used, the time of day, the specific location and any other relevant information, such as weather conditions.

☐ Assess the viability of your methods.

☐ Use diagrams and/or photos to demonstrate your methods.

Criterion C: Quality and treatment of information collected*

Marks available: 5 marks

This criterion assesses the usefulness of the data collected and the way you present this information in your report. When displaying data in your report, refer to the following points:

☐ The techniques you use to display your data will vary depending on the data, but they may include statistical tests (such as confidence levels), graphs, diagrams, maps, annotated photos and images, matrices and field sketches. Think about which of these techniques (or others) would be most appropriate to display your findings.

☐ Ensure that your data is positioned logically within your written analysis (*criterion D*).

☐ Think about the numbering, labelling and annotating of tables, graphs and diagrams; follow an accepted convention and ensure the presentation of these aspects is consistent.

*Criterion C assesses information display and does not include a word count. However, lengthy annotations of your diagrams may be factored into the word count so aim to keep these brief.

Criterion D: Written analysis

Marks available: 8 marks
Suggested word count: 1,350 words

For this criterion, you should demonstrate your knowledge and understanding by interpreting and explaining the data collected. The presentation of your data (*criterion C*) should be integrated within your written analysis. Refer to the following checklist when writing this section:

☐ In your analysis of the data, always relate the results to the central themes addressed in the fieldwork question and geographic context of the study; do the data help to answer the fieldwork question?

☐ Identify any spatial patterns and trends in your data, and explain these patterns.

☐ Attempt to explain any anomalies in your data, possibly with reference to secondary sources.

☐ Ensure your data are displayed effectively (*criterion C*) within your written analysis to support your points.

Criterion E: Conclusion

Marks available: 2 marks
Suggested word count: 200 words

In your conclusion, you should summarize the findings of your fieldwork investigation; refer to the following points:

☐ There must be a clear, concise statement that answers the fieldwork question.

☐ Refer to the data of your investigation to support your conclusion.

☐ Compare the results of the investigation against your initial hypotheses; it is acceptable for the conclusion to state that the findings do not match any of your preliminary judgements or projections.

Criterion F: Evaluation

Marks available: 3 marks
Suggested word count: 300 words

In your evaluation, you should review your investigation. In particular, you should think about the following aspects:

☐ Thoroughly evaluate your methods of collecting primary data in the field. Assess how you could improve these methods and the quality of data achieved for future investigations.

☐ Consider the factors that may have affected the validity of data; for example, these factors could include personal bias or unpredicted external circumstances such as the weather.

☐ Suggest viable and realistic ways in which the study as whole might be improved in the future. Could the fieldwork question and sites be improved?

☐ Suggest possible extensions to your investigation.

You should note that these criteria can be represented by material in any part of the report and assessment of the criteria is not confined to one section.

Report checklist

Complete this checklist before you submit your fieldwork report.

Tasks	Completed
The work is within the 2,500 word limit.	
There is a title page with the candidate name and number.	
There is a contents page.	
All the pages are numbered.	
All illustrations have figure numbers.	
All illustrations are close to the relevant text.	
All sources are correctly referenced.	
The appendix contains only raw information.	
The report has a fieldwork question.	
All methods of information collection are fully justified.	
All maps have normal conventions of title, scale, north point and key.	
The analysis refers to the fieldwork question and the information collected.	
There is a conclusion.	
The evaluation makes recommendations for improvements.	

PRACTICE EXAM PAPERS

At this point, you will have re-familiarized yourself with the content from the options and units of the IB Geography syllabus. Additionally, you will have picked up some key techniques and skills to refine your exam approach. It is now time to put these skills to the test; in this section you will find practice examination papers, **1, 2 and 3**, with the same structure as the external assessment you will complete at the end of the DP course. Paper 3 is completed by higher level students only. Answers to these papers are available at **www.oxfordsecondary.com/ib-prepared-support**.

Paper 1 (SL and HL)

Option A: Freshwater

1. Examine the diagram below showing the Lake Ontario drainage basin.

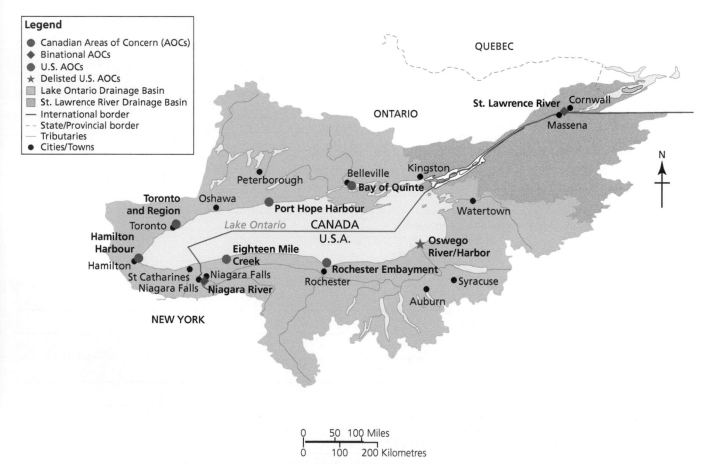

Legend
- ● Canadian Areas of Concern (AOCs)
- ◆ Binational AOCs
- ● U.S. AOCs
- ★ Delisted U.S. AOCs
- ▨ Lake Ontario Drainage Basin
- ▨ St. Lawrence River Drainage Basin
- — International border
- -- State/Provincial border
- — Tributaries
- ● Cities/Towns

Lake Ontario drainage basin

(a) Estimate the width of the drainage basin at its widest point. [1]

(b) Using evidence from the map, explain why a drainage basin is a categorised as an 'open system'. [1]

(c) Explain how processes of erosion can be affected by one spatial factor and one temporal factor. [2+2]

(d) With reference to an example that you have studied, explain why wetlands should be protected. [4]

Either

2. Examine how human influences in one part of a drainage basin can cause create change downstream in a drainage basin. [10]

Or

3. Examine the importance of an integrated basin management plan that you have studied. [10]

Option B: Oceans

4. Examine the photo on the right, which shows a coastline at Bald Head Island, North Carolina.

   (a) Identify two coastal landforms shown in the photo. [2]

   (b) Using an annotated diagram, explain how the landform identified in question (a) has been formed. [4]

   (c) Suggest how two different types of land-use can create pressures on coastlines such as this. [2+2]

Either

5. Examine the economic and ecological value of coastal margins. [10]

Or

6. Evaluate initiatives that have been implemented to manage ocean pollution. [10]

Coastline at Bald Head Island, North Carolina

Option C: Extreme Environments

7. Study the map below.

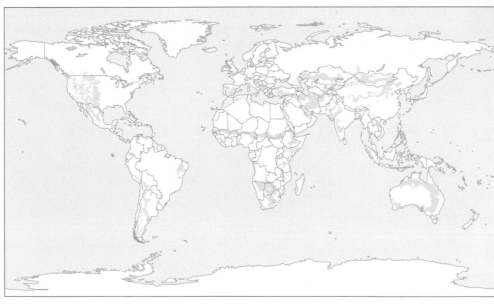

Map showing semi-arid environments in yellow

   (a) Describe the distribution of semi-arid climates using the map above. [2]

   (b) Outline two challenges for human habitation in an extreme environment. [2+2]

   (c) Explain how arid environments can be utilized for economic benefits. [2+2]

Either

8. Examine the interrelationship between global climate change and extreme environments. [10]

Or

9. Examine the conflict between different stakeholders in one extreme environment that you have studied. [10]

Option D: Geophysical hazards

10. Study the map on the left, which shows the level of vulnerability in relation to liquefaction in the district of Gisborne in New Zealand.

    (a) Using map evidence, describe the distribution of high vulnerability in the district. [2]
    (b) Examine the photo on the left:
        Suggest two reasons why the building in the image collapsed. [2+2]
    (c) Explain one strength and one weakness for a scale that measures the magnitude of an earthquake or a volcanic eruption. [4]

Either

11. Examine how the economic and social development of a place can influence geophysical hazard risks. [10]

Or

12. Examine how planning and pre-event strategies ensure that places are able to cope with geophysical hazard events. [10]

Option E: Leisure, sport and tourism

13. Study the chart below which shows the number of passengers passing through the biggest airports in Thailand, in relation to the number of passengers they are designed to handle.

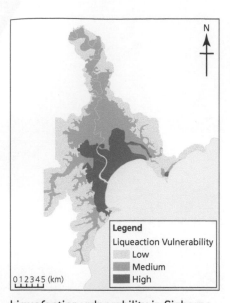

Liquefaction vulnerability in Gisborne, New Zealand
Source of data: Gisborne District Council

Collapsed building in the Belice valley, Sicily

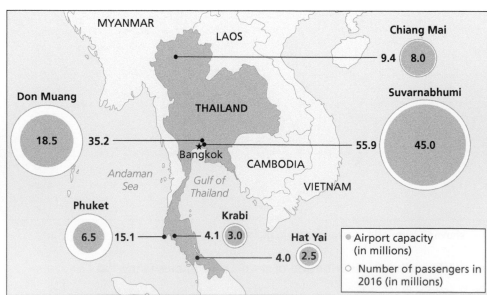

The capacity of Thailand's airports and number of passengers in 2016
Source of data: Airports of Thailand Pcl; Civil Aviation Authority of Thailand

(a) (i) State the airport with the largest amount of
    overcapacity. [1]
  (ii) State the airport with the smallest amount of
    overcapacity. [1]
(b) Outline one way in which tourism congestion (people)
    can be managed. [2]
(c) Using an example that you have studied, explain
    the economic, social and environmental benefits of
    sustainable tourism to a low-income country. [2+2+2]

14. Examine the influence of political decisions and social
    movements that have affected the participation of different
    groups in international sports and sporting events. [10]

Or

15. Examine how primary touristic resources can cause the growth of
    tourism hotspots. [10]

Option F: Food and health

16. Study the schematic below.

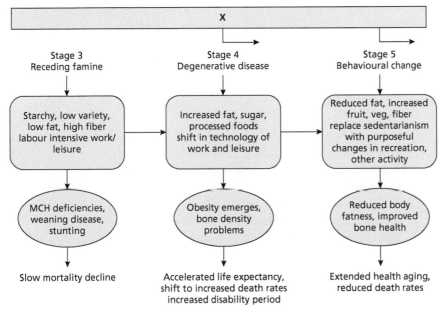

Schematic showing stages of nutrition transition

(a) State two factors that could be included in the box
    labelled 'X'. [2]
(b) Describe what is meant by the term 'epidemiological
    transition'. [2]
(c) Explain one advantage for each of the following:
    • genetically modified organisms (GMOs)
    • vertical farming
    • in vitro meat in improving food production [2+2+2]

Either

17. Examine the geographic factors that caused the diffusion
    of one vector-borne disease and one water-borne disease. [10]

Or

18. Evaluate the strategies designed to overcome famine. [10]

Option G: Urban environments

19. Study the chart below.

Drivers' time spent in peak traffic congestion

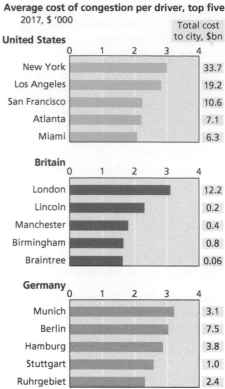

Average cost of congestion per driver, top five

Congestion in cities, 2017
Source of data: INRIX Research

(a) Compare the cost of congestion to top 5 cities in the USA and Britain. [1+1]

(b) Outline two ways in which traffic congestion can be reduced. [2+2]

(c) With reference to an example that you have studied, explain two factors that have resulted in a depletion of green space. [2+2]

Either

20. Examine the extent to which an eco-city design can impact the urban ecological footprint. [10]

Or

21. Examine the links between deindustrialization and the cycle of deprivation in urban environments. [10]

Paper 2 (SL and HL)

Part A

Unit 1: Changing population

1. Examine the map below, which shows the population distribution for Saudi Arabia:

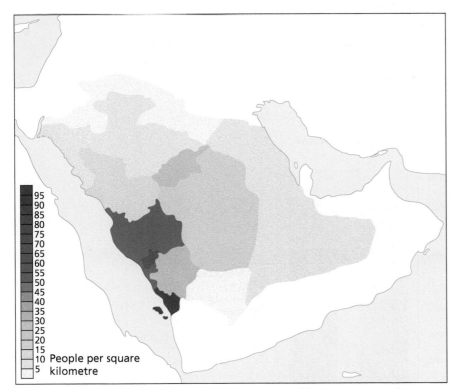

Population density in Saudi Arabia
Source of data: Socioeconomic Data and Applications Center

(a) Describe the population distribution.	[2]
(b) Suggest one physical reason for an area that contains a low population density.	[2]
(c) With reference to a megacity that you have studied, explain two reasons for its growth.	[3+3]

Unit 2: Global climate—vulnerability and resilience

2. Study the diagram below which shows the carbon dioxide emissions for countries and regions between 2012–2040.

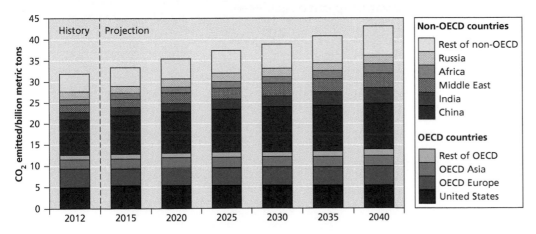

Projected energy-related carbon dioxide emissions by country or region (2012–2040)
Source of data: US Energy Information Administration (2016)

(a) (i) In 2020, estimate the difference between the OECD and non-OECD countries in terms of the amount of carbon dioxide emitted. [1]
  (ii) Identify the country that is projected to have the highest increase (non-proportional) between 2015 and 2040. [1]
(b) (i) Define the term "positive feedback loop". [1]
  (ii) Explain the operation of **one** positive feedback loop that contributes to climate change. [3]
(c) Explain **one** positive effect and **one** negative effect from global climate change upon people in low-income countries. [2+2]

Unit 3: Global resource consumption and security

3. Study the graph below.

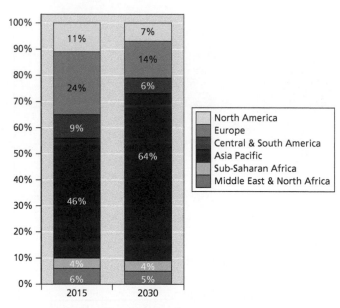

Global middle class population, % of total
Source of data: Brookings Institution (2017)

(a) Identify the region that will experience the largest change between 2015 and 2030. [1]
(b) Referring to regions with different levels of economic development, describe the changes between 2015 and 2030. [3]
(c) Suggest how strategies designed to achieve resource stewardship can be achieved. [3+3]

Part B

4. Study the infographic below:

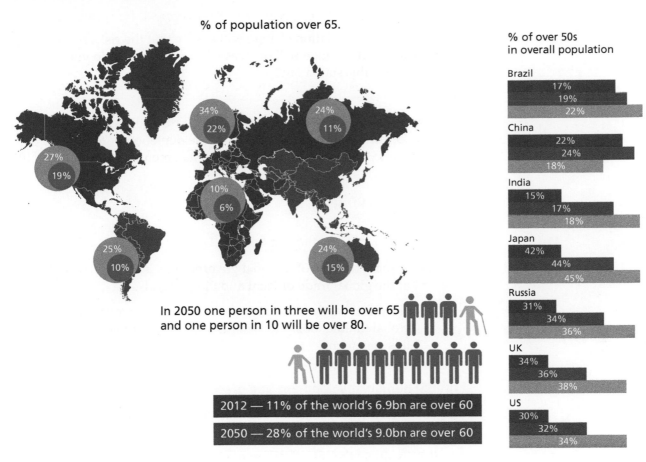

% of population over 65.

% of over 50s in overall population

In 2050 one person in three will be over 65 and one person in 10 will be over 80.

2012 — 11% of the world's 6.9bn are over 60

2050 — 28% of the world's 9.0bn are over 60

Brazil 17% 19% 22%
China 22% 24% 18%
India 15% 17% 18%
Japan 42% 44% 45%
Russia 31% 34% 36%
UK 34% 36% 38%
US 30% 32% 34%

The World's ageing population

(a) (i) Identify the continent will experience the largest increase in its percentage of people over the age of 65. [1]

(ii) Identify the continent will experience the smallest increase in its percentage of people over the age of 65. [1]

(b) (i) State the number of people who will be over the age of 65 in the year 2050. [1]

(ii) State the number of people who will be over the age of 80 in the year 2050. [1]

(c) Suggest 3 ways in which the infographic could be improved. [2+2+2]

Part C

Either

5. "Managing population change will ensure that the risks from global climate change will decrease." To what extent do you agree with this statement? [10]

Or

6. "Despite the impacts from global climate change, global development is not possible without the depletion of resources." To what extent do you agree with this statement? [10]

Paper 3 (HL only)

Unit 4: Power, places and networks

1. (a) Explain the role of national governments in the movement of the global trade of legal and illegal goods. [12]

   (b) "Physical barriers no longer have relevance in relation to global interactions." To what extent do you agree with this statement? [16]

Unit 5: Human development and diversity

2. (a) Analyse the processes which affect cultural diversity at a global scale. [12]

   (b) Evaluate the role of different stakeholders involved in the global processes of human development. [16]

Unit 6: Global risks and resilience

3. (a) Explain how the movement of manufacturing industry affects people and places. [12]

   (b) "Increasing resilience to global risks can only increase global interactions." To what extent is this statement true? [16]

Index